Transactions Of The Institution Of Engineers And Shipbuilders In Scotland

Yours very sincerely,

Wm Denny

TRANSACTIONS

OF

𝕴𝖓𝖘𝖙𝖎𝖙𝖚𝖙𝖎𝖔𝖓 𝖔𝖋 𝕰𝖓𝖌𝖎𝖓𝖊𝖊𝖗𝖘 𝖆𝖓𝖉 𝕾𝖍𝖎𝖕𝖇𝖚𝖎𝖑𝖉𝖊𝖗𝖘

IN SCOTLAND

(INCORPORATED).

VOLUME XXX.

THIRTIETH SESSION,
1886-87.

EDITED BY THE SECRETARY.

GLASGOW:
PRINTED FOR THE INSTITUTION BY
WILLIAM MUNRO, 80 GORDON STREET.
1887.

OFFICE-BEARERS.

THIRTIETH SESSION, 1886-87.

President.
WILLIAM DENNY, F.R.S.E.

Vice-Presidents.
CHARLES P. HOGG. | JOHN INGLIS, Jun.
ROBERT DUNDAS.

Councillors.
FRANCIS ELGAR, LL.D.   ALEXANDER C. KIRK.
CHAS. C. LINDSAY.   GEORGE L. WATSON.
JOHN THOMSON.   ALEXANDER SIMPSON.
HENRY DYER, M.A.   W. RENNY WATSON.
GEORGE RUSSELL.   WILLIAM FOULIS.

Finance Committee.
WILLIAM DENNY, Convener.
ALEXANDER SIMPSON.
W. RENNY WATSON.
ROBERT DUNDAS.

Library Committee.
C. C. LINDSAY, Convener
CHARLES P. HOGG.
HENRY DYER.
GEORGE L. WATSON.

Honorary Councillors—Past Presidents.
WALTER M. NEILSON.   ROBERT DUNCAN.
JAMES G. LAWRIE.   HAZELTON R. ROBSON.
JAMES M. GALE.   ROBERT MANSEL.
DAVID ROWAN.   JAMES REID.
PROF. JAMES THOMSON, LL.D., F.R.S.

Honorary Treasurer.
JAMES M. GALE, 23 Miller Street.

Secretary and Editor of Transactions.
W. J. MILLAR, 261 West George Street.

Honorary Librarian.
C. C. LINDSAY, 167 St. Vincent Street.

Sub-Librarian.
THOMAS NAPIER, Institution Rooms, 207 Bath Street.

995112

CONTENTS.

THIRTIETH SESSION, 1886-87.

PLATES.

Premiums Awarded

PAPERS READ DURING SESSION 1885-86.

A PREMIUM OF BOOKS

To Mr Henry Dyer, C.E., M.A., for his paper on "The Present State of the Theory of the Steam Engine, and some of its bearings on current Marine Engineering Practice."

A PREMIUM OF BOOKS

To Mr Andrew S. Biggart, C.E., for his paper on "The Forth Bridge Great Caissons, their Structure, Building, and Founding."

INSTITUTION of ENGINEERS & SHIPBUILDERS

IN SCOTLAND.

(INCORPORATED.)

THIRTIETH SESSION, 1886-87.

October 26th, 1886.   Mr CHARLES PULLAR HOGG, Vice-President,
in the Chair.

Mr C. P. HOGG, Vice President, said, in opening the proceedings, that he occupied the chair on that occasion owing to the absence of Mr William Denny, the President, who was now, he believed, on his way home from South America. He was not quite sure that Mr Denny could be at the next meeting, but they might depend upon having him present before Christmas, to give the inaugural address.

On the Safety Governor.

By Mr JAMES W. MACFARLANE.

(SEE PLATES I. AND II.)

Received 22nd, and Read 26th October, 1886.

THE steam engine governor is a subject which has received a large amount of attention from mechanical engineers and others, if one may judge from the number of patents which have been taken out for real or imagined improvements on it; but till now there has not, so far as I am aware, been any paper bearing on the subject before this Institution.

Under these circumstances, I take the liberty of bringing before you a governor which I venture to think may have some features worthy of your consideration.

For reasons which will afterwards be apparent, this governor has been named "The Safety Governor."

It will be observed that there is a centrifugal part, A (see Plate I., Fig. 1), contained in a steam chest, B, having a flanged branch, B_1, connected to the steam pipe from the boiler, and a flanged branch, B_2, connected to branch pipe on steam engine cylinder, or steam pipe leading thereto.

Communication between these parts being opened or closed as desired by a stop valve, K, placed in the right hand part of the chest, B, and operated by a hand wheel, Q, secured on a screwed spindle, Q_1, connected to the stop valve, K, by a screwed collar, K_1, and packed steam-tight by the stuffing box and gland, K_2, as shown. To show the position of the stop valve, a pointer, R, screwed into the end of the steam chest, B, is provided, in conjunction with the hand wheel, Q, the said pointer being graduated to same pitch as screw of stop valve spindle, Q_1.

The stop valve, K, which when open admits steam into the annulus, B_2, surrounding it, is provided with valve seat, O, and guide liner, P, through which the sleeve, L, on back of stop valve, K, slides easily. The said sleeve, L, is formed with circumferential rows of port holes, M, and in the interior part of it the throttle valve, C, is contained. The throttle valve (having also circumferential ports, N, similar to those in the stop valve), is attached to a spindle, G, which turns freely in the centre of a sliding block, F.

Longitudinal motion is communicated from the hemispherical pendulums, D, to the throttle valve, C, by means of the bell crank levers, D_1, centred on pins, E, in a crosshead, I_2, and actuating the sliding block, F, to which is attached the throttle valve spindle, G, by means of its collar, G_1.

Rotary motion is communicated from the driving pulley, X, to the pendulums, D, by means of the revolving spindle, I, passing through a stuffing box, T, into the steam chest, B, and having a crosshead, I_2, on its inner end, to which the pendulums, D, are attached, and which turns in the liners, V, W, and stuffing gland, U, shown.

In a chamber, I_1, of the revolving spindle, I, a helical spring, H, is placed which presses against the sliding block, F, and thereby counteracts the centrifugal action of the two hemispherical pendulums, D.

A small spindle, J, passes through the longitudinal centre of the revolving spindle, I, for the purpose of adjusting the spring, H, and has a screw and jam nut on the outer end, and a cap piece, J_1, on inner end pressing against the spring, H.

The whole of the working parts of the governor are carried by the cover, S, of the steam chest, B, the two latter parts being connected by bolts or screws, making a steam-tight joint.

A lubricator, Y, is provided for lubricating the working parts of the governor, after which, the lubricating material passes with the steam into the steam cylinder of the engine.

Having now described briefly the construction of this governor, I may here be allowed to state in general terms what may be considered as the essential points to be aimed at in the designing of a good steam engine governor. They are as follows, viz. :—

1st, Safety ; 2nd, Sensitiveness ; 3rd, Simplicity ;

and, under these headings, I will now endeavour to explain how far these important points have been attained in the governor which is the subject of the present paper.

1st, *Safety.*—There have been many accidents recorded, resulting in loss of life and property, through engines running away ; more especially has this been the case with engines used for driving grinding and centrifugal machinery, and also recently in engines for electric lighting purposes.

These accidents have generally been traced to the governor, the most frequent source of danger being due to the governor belt breaking or slipping off the pulley, and consequently numerous attempts have been made to rectify this defect. The objections to the more common devices for this purpose have been that they require some extra attention from the attendant, and also on account of the extra cost. In the governor before you the stop valve is formed into an adjustable seat for the throttle valve. When the stop valve is opened the engine begins to work, and the throttle valve to shut the ports in the stop valve ; the stop valve is opened until the hand wheel comes against the check on the pointer, the throttle valve gradually following up until the engine arrives at its proper speed, the initial and final relative positions being as shown in Figs. 3 and 4 (see Plate II.).

The engine is now working at its normal speed, and if the governor belt now breaks or slips off, the governor stops revolving, and the centrifugal parts immediately resume their original position. The stop valve, however, remains open, and we get the new relative position of the stop and throttle valves as shown in Fig. 5. The steam being shut off, the engine is brought to rest.

You will observe that the attendant has not had anything to do beyond opening the stop valve, the automatic stopping arrangement being entirely beyond his control.

It will be explained further on that this governor does not admit of a greater deviation in speed than from 10 per cent. to 12 per cent., when the stop valve is full open, and the hand wheel against the stop check on the pointer. This is a feature of some importance where an engine is liable to a sudden extra strain, as may happen often with some engines, say in the case of engines working dredgers. Any extra strain which would decrease the speed beyond the minimum allowed, would have the effect of closing the ports the backward way, rapidly reducing the speed, and finally stopping the engine.

Practically the same thing would occur should the steam pressure fall so much that it would not be sufficient to keep up the speed of the engine to the minimum.

But if it were desired to continue working the engine, the stop valve would require to be partly drawn back, as in Fig. 8, when the engine would continue working so long as the steam was able to turn it.

In some cases, the total stoppage of the engine caused by a sudden extra strain might be a disadvantage, and to provide against this, the position of the stop valve can be so arranged that the throttle valve could not shut off the steam wholly the backward way, but still have the port slightly open, sufficient to keep the engine moving until the extra strain was removed, when the engine would immediately recover its normal speed. In this case the safety arrangement would, if an accident should happen, not stop the engine totally, but cause the speed to be considerably reduced. This position of the stop valve can easily be found for each engine experimentally, by throwing off the governor belt when the engine is working light with the maximum pressure of steam, and causing it to stop, and then drawing back the stop valve a small way, until the engine gets steam to keep it working at a reduced speed, and making the new position of the stop valve the limit of the opening in usual work.

2nd, *Sensitiveness.*—In governors of this class a constant source of trouble is the stuffing box between the centrifugal part and the throttle valve. The stuffing box is either kept so slack that there is a continual leakage, or it is tightened up so much to prevent leakage that the sensitiveness of the governor is very considerably impaired; and in the hands of an unskilled or careless attendant, governors of this kind become practically of little account. Various contrivances have been tried to get over this difficulty, some of which have been successful only by additional complication at the expense of their simplicity, and also being necessarily less sensitive on account of the extra joints introduced, each joint adding not only friction, but a certain amount of slack which has to be taken up by extra travel of the governor before it can act on the throttle valve.

In the governor before you this difficulty has been largely overcome by placing the centrifugal part in the steam, and which is then in direct communication with the throttle valve without the intervention of rods, levers, or joints, and without having to pass through any stuffing box. A considerable amount of variable friction is therefore got rid of, and also the sometimes doubtful care of the attendant.

A not inconsiderable advantage to be derived from this reduction in friction is that the centrifugal part can be made smaller relatively to the size of the throttle valve, a circumstance which enables the governor to be better adapted for high speeds.

There are, however, other things besides the friction which affect the sensitiveness of a governor, and I shall now endeavour to show how these points have been considered in the present instance, and also how far they have been borne out in practice.

The elasticity of a helical spring is represented by the formula

$$L = \left(\frac{cd^4}{nr^3}\right) E^{*} \quad \dots\dots\dots\dots\dots(1)$$

from which it will be observed that the quantity within the brackets is a constant for any particular spring, and that the quantities L and E are variables depending upon each other, hence the equation (1) may be written,

$$y = ax,$$

which is the equation to a straight line referred to rectangular axes, and passing through the origin.

Let Ox, Oy, Fig. 9 (see Plate II.), be the rectangular axes, make ON = E and NP = L at right angles to Ox, the straight line joining O and P will represent a spring whose load is NP and compression ON.

Now, the load on the spring in this governor is the centrifugal force of the hemispherical pendulums acting through the bell-crank levers, and the centrifugal force of a revolving body is

$$= f = \frac{Mv^2}{R},$$

where $v^2 = (2\pi RN)^2$, and $M = \frac{W}{g}$;

$$\therefore f = \frac{4W\pi^2RN^2}{g},$$

and putting C = the constant $\frac{4\pi^2}{g}$, we have

$$f = CWRN^2 \quad \dots\dots\dots\dots\dots\dots (2)$$

It will now be necessary, before proceeding further, to observe that the centres of the pendulum bell-crank levers are so placed that

---

* L = load in lbs.
  E — compression in inches.
  n — number of turns of coil effective.
  r — mean radius of coil in inches.
  d — diameter of steel wire in inches.
    — side of square steel wire in inches
  c — 187,500 for round steel.
    — 256,818 for square steel.

the difference between the maximum and minimum radii of the centres of gravity of the pendulum has to the total travel of the throttle valve a nearly constant ratio, and is approximately equal to the ratio of the lengths of the bell-crank levers, and may be represented thus,

$$\frac{R - h}{T} = B = \text{a constant} \dots\dots\dots\dots(3)$$

h being the minimum radius = the height of the centre of gravity of the hemispherical pendulum above the base.

Of course this may be easily seen by the diagrams, Figs. 3 to 8 (Plate II.). The centrifugal force and the load on the spring acting in lines (in the same plane) at right angles to each other, if we make the centre lines of the bell-crank levers also at right angles to each other, the ratio of the leverages will then be perfectly constant. For practical reasons the levers are not quite at right angles to each other, but are so near that the quantity B may be taken as constant without sensible error.

Now, the load on the spring is equal to the product of the centrifugal force of the two hemispherical pendulums and the ratio of the bell-crank levers, consequently,

$$L = Bf = BCWRN^2 \dots\dots\dots\dots(4)$$

but the radius $R = h + BT$, from equation 3, therefore

$$L = BCW (h + BT) N^2 \dots\dots\dots\dots(5)$$

the general equation for any travel t of the throttle valve being

$$L_t = BCW (h + BT_t) N^2_t \dots\dots\dots\dots(6)$$

As the hemispherical pendulum, except in its initial position, is not symmetrical with respect to the plane perpendicular to the axis of revolution, and passing through the centre of gravity, it follows that the equation $L = BCWRN^2$ is not mathematically accurate for positions of the pendulum other than the initial position, but is so near that in taking an actual example, the revolutions do not vary (between the maximum and minimum positions of throttle corresponding to the width of port, Figs. 6 and 7, Plate II.) from the true amount more than one revolution in four hundred, and that under the most unfavourable conditions.[*]

Measure, to any convenient scale, diagram Fig. 9 (see Plate II.).

$ON = T_s = $ travel of the throttle corresponding to port full closed as in Fig. 6.

$OQ = T_2 = $ travel of the throttle corresponding to port full open as in Fig. 7.

$QN = $ width of port $= T_s - T_2 = \overline{ON} - \overline{OQ}$.

Now, if we assume N_s to be the maximum number of revolutions corresponding to the travel T_s of the throttle valve, the revolutions for any less travel of the throttle will either be equal to N_s, or less. Measure to the same or any other convenient scale.

$NP = L_s = BCW (h + BT_s) N_s^2 = $ load on spring corresponding to travel of throttle $T_s = ON$,

but when the travel of the throttle valve $= T_1 = 0$, we have

$$L_1 = BCW (h) N_1^2 \quad\dots\dots\dots\dots\dots\dots(7)$$

Now, N_1 may be either equal to N_s or less; suppose it is equal.

$$L_1 = BCW (h) N_s^2 \quad\dots\dots\dots\dots\dots\dots(8)$$

Measure OU on the axis of $y = L_1$, and join P.U. and produce to cut the axis of x in V.

Again, if $N_1 = 0$, L_1 becomes $= 0$, consequently the length of OU will vary according to the speed, from

$$L_1 = BCWhN_s^2 \quad \text{to} \quad L_1 = 0.$$

We have here two springs represented by the lines PO and PV, either of which will answer the conditions—viz., that at the travel

---

[*] The actual calculated load on the spring due to the centrifugal force of the two hemispherical pendulums (neglecting the mass of the levers) is

$$L = \frac{2Mw^2 \cos C}{d \cos B}\left[\left(cd + hd - ch - \frac{b^2}{2}\right) \sin C + n (c + h) \cos C\right].$$

where $h = \dfrac{3}{4\pi}\left\{ b \dfrac{a^2 - 2b^2}{a^3 - b^3} + \dfrac{a^4}{(a^2 - b^2)^{\frac{1}{2}}} \cos^{-1}\dfrac{b}{a}\right\}$

and $2M = \dfrac{W}{g} = \dfrac{4\pi}{3g} (a^2 - b^2)^{\frac{1}{2}} \times k$ for kind of material.

(See diagram, Fig. 10, Plate II.)

T_s of the throttle valve the speed shall be N_s—but with respect to the spring PV, it is the weakest that can be employed, and is such that the speed of the governor is constant; in other words, by putting a spring into the governor whose elasticity = VN = E, the throttle will travel full out or full in with the smallest increase or decrease of the speed. This, you will observe, is the particular property of the parabolic governor, and it is obtained with the simplest form of pendulum, and only one spring.

Again, the spring represented by PO is the strongest that can be employed, and is such that the speed of the governor will vary between N_s and 0 within the same limits of travel of the throttle— viz., T_s to O, or $\overline{ON}$ to O—but as only a portion of the travel of the throttle is required in this governor for closing and opening the port, the range of the travel of the throttle we have to consider is that part represented by QN = the width of the port. Draw QS parallel to NP, then by the property of similar triangles,

$$QS : NP :: OQ : \acute{O}N ;$$

that is, as

$$L_2 : L_2 :: E_2 : E_3,$$

from which we find

$$L_2 = \frac{L_3 E_2}{E_3} \quad\dots\dots\dots\dots\dots(9)$$

But $L_2 = BCW\,(h + BT_2)\,N_2^2$, from equation (6), and

$$N_2^2 = \frac{L_2}{BCW\,(h + BT_2)} \quad\dots\dots\dots\dots(10)$$

and by replacing L_2 by its equivalent in (9), we have

$$N_2^2 = \frac{L_3 E_2}{E_3 BCW\,(h + BT_2)}\dots\dots\dots\dots(11)$$

from which we find that if the maximum speed corresponding to port full closed, as in Fig. 6, = N_s, the minimum speed corresponding to port full open, as in Fig. 7, will be N_2, the reduction in speed from N_s to N_2 in any particular case not being more than 12 per cent., and that with the least sensitive spring.

Seeing now that we have fixed the limits of deviation, it is open to us to employ a spring that will give us any desired degree of sensitiveness between the specified limits. Suppose we fix n as the desired minimum speed, corresponding to N as a maximum; from equation (6) we find

$$l = BCW\ (h + BT_2)\ n^2 \dots\dots\dots\dots(12)$$

Make QW = l, and join PW and produce to meet the axis of x in X. PX will represent the required spring to give the desired deviation in speed at the limits specified; and

NX = E = the elasticity of the spring;

NP = L = the corresponding load on the spring;

OX = the initial compression; that is,

the amount the spring is tightened up before the centrifugal force begins to act upon it.

As a perfectly isochronous governor is liable to allow the engine to hunt, it is found better to fit these governors with a spring that is not too sensitive, as we have a means at our disposal of making it as sensitive as we choose, by taking advantage of the movable seat of the throttle valve.

Suppose the governor is fitted with the spring represented by PX (Fig. 9, Plate II.), having the initial compression = OX; let this initial compression be increased (by tightening up the spring) to OV, the limiting amount; draw VP_1 parallel to XP, and PZ parallel to ON; the speed of the engine will be increased to an amount corresponding to the increased load represented by NP_1; but by drawing back the stop valve until the throttle valve travel is reduced a distance = PZ, the speed of the governor will be decreased by the same amount that it was increased by tightening up the spring, and consequently the governor is rendered isochronous without changing the speed of the engine, and also without changing the spring.

Again, suppose it were desirable to increase the speed of the engine permanently, this is done by tightening up the spring as previously explained $\left(\text{the limit of initial compression being OV} = \dfrac{h}{B}.\right)$ Draw VP_2 parallel to OP, the minimum and maximum loads on the spring due to the extreme speeds will be represented by the lengths of the lines NP, NP$_2$ respectively, and the actual speeds by the square roots of these lengths, now by the property of similar triangles already referred to,

$$\text{ON} : \text{VN} : \text{NP} : \text{NP}_2 ;$$

and as the lengths ON · VN are known for each governor, we find the extreme speeds to be respectively as

$$\sqrt{\overline{\text{ON}}} : \sqrt{\overline{\text{VN}}}.$$

If N$_s$ is the speed corresponding to the load represented by NP, the speed corresponding to the load NP$_2$ will be $= \text{N}_s \dfrac{\sqrt{\overline{\text{VN}}}}{\sqrt{\overline{\text{ON}}}}$, and the difference or increase in speed $= \text{N}_s \dfrac{\sqrt{\overline{\text{VN}}}}{\sqrt{\overline{\text{ON}}}} - \text{N}_s$, which is found to be limited to about 36 per cent. in most cases.

A considerable number of tests have been made of the working of a 3-inch governor fitted on an engine having a pair of 12-inch cylinders, and which is used for driving the machinery and cranes in an engineering establishment, with pressures of steam varying from 62 to 35 lbs. per square inch on the gauge. The speeds have been found to vary from a maximum of 287 revolutions to a minimum of 279 revolutions per minute, or a total variation of 3 per cent., some of the tests being taken while the engine was running light, and some while doing the usual work of the shop. Knowing from actual test the strength of the spring in use, also the amount of the initial compression, we find by equation (10) that the governor has a maximum permissible deviation of about 10 per cent., which can be easily verified practically in the following manner—viz., from the average of a number of tests, the speed when the stop-valve is full open is found to be 283 revolutions per minute : by drawing back the stop-valve a distance equal to the width of the port (the effect of which is to reduce the travel of the throttle from N to Q, Fig. 9, Plate II.), we find the average speed to be 255 revolutions per minute—a reduction of 28 revolutions, or about 10 per cent., the actual and calculated amounts almost exactly agreeing with each other. It will be observed that the maximum variation of 3 per cent. is obtained under very wide conditions of steam pressure and load on the engine. Were these conditions more uniform, the total deviation would, no doubt, be less. It is also to be borne in mind that the spring employed is very nearly the stiffest and least sensitive, and we have it within our power to make this governor still more sensitive by one or other of the methods already referred to.

I learn from good authority that the deviations allowed by ordinary governors vary from 5 to 15 per cent., and that the deviations due to friction alone, in what are considered to be good governors, vary from $\frac{1}{2}$ to 6 per cent. I cannot from personal experience corroborate these statements, but assuming that they are correct, you will observe that the deviations just recorded for the governor under consideration are very considerably below these figures.

3rd, *Simplicity.*—The small governor which I have brought here to-night for your inspection has a $1\frac{1}{4}$-inch bore of steam pipe, and is suitable for an engine with steam cylinder 5-inch to 7-inch diameter. It is one of a number of the same size made and used by Messrs. Watson, Laidlaw, & Co., of this city, on small diagonal engines, for driving centrifugal machines, and for other purposes.

(The other two governors lying on the table, having respectively $1\frac{1}{2}$-inch and 2-inch bore of steam pipe, were sent here to-day by Messrs. Steven & Struthers, brassfounders in Elliot Street, and they may be taken as fair samples of a lot of governors which have recently been made by them.)

If you take this governor apart you will find that there are, all told, without including the lubricator (which is an accessory to the engine apart altogether from the governor), 32 parts or pieces, including even the screws and split pins. A few years ago I had occasion to take apart a well-known governor of this class, but without any of the special advantages claimed for the governor before you, and it had not less than 64 distinct pieces, being exactly double the number mentioned above. Of course, in the larger sizes of governor, such as is represented by the drawings, there are more pieces, but the total number does not exceed 42.

The construction of the various details is so arranged that the parts may be easily taken asunder, it being only the work of a few minutes to take the whole thing apart and put it together again, and the pieces are of sufficient size that even in the small governor here they are quite within the capabilities of an ordinary engineering shop to make or to repair.

As some doubts have been expressed as to the efficient lubrication of the governor, I may here be allowed to say a word or two on this point. For lubricating the internal parts a suet lubricator drops the oil into a cup formed in the centre boss of the steam-chest cover, and from thence it finds its way by gravity to the revolving spindle bearing, and into the chamber containing the helical spring, and lubricates the sliding piece in the cross-head, and is afterwards thrown out by centrifugal action to the joint-pins, ultimately finding its way to the steam-engine cylinder. In fact, the whole lubrication required for the cylinder may be made to pass through the governor, thereby keeping it practically running in oil. Now, the oil will find its way into the centre of the spindle at all speeds below a certain maximum limit, that limit being when the velocity of the circumference of the driving spindle is equal to the velocity acquired by a falling body falling a distance equal to the height of the top of the cup to the centre of the spindle. Putting this in the form of an equation, we have

$$n = \frac{\sqrt{2g\mathrm{H}}}{\pi d}.$$

In the case of the $1\frac{1}{4}$-inch governor, the limiting speed from the formula is 975 revolutions per minute.

Advantage can be taken of the direction of rotation to slant the oil hole (which passes through the spindle) forward in a right hand direction, giving the oil what may be called an inward flow, materially assisting the action of gravity.

The bearing for the outer end of the driving spindle is supplied with oil from the cup surrounding the stuffing gland, the oil working its way to the outer end of the bearing drops into the groove formed on the inside of the pulley, and is carried round by reason of the cohesion due to the centrifugal action. A small bent tube, one end of which is in the stuffing-box cup, and the other end pressing against the rim of the pulley, catches the oil collected there as the pulley revolves and returns it to the cup, forming what might be called a continuous lubricator.

In answer to questions by Mr J. M. GALE,

Mr MACFARLANE said that the governor had not been specially got up for use in screw steam boats; it had not yet been tried for that purpose. It had been working for fifteen or sixteen months.

The CHAIRMAN said he supposed they would be glad to see the paper in print before making any remarks upon the governor; and as the rules of the Institution required that the discussion be adjourned till a second night, it would now be continued to next meeting.

The discussion on this paper was resumed on the 23rd November, 1886; when

Mr GEORGE RUSSELL said he had read the paper with much interest, and had found that it was a very good paper. Mr Macfarlane observed in his introduction that no paper on the subject had been previously brought before this Institution; but he had overlooked various papers in the earlier volumes of Transactions. Mr Macfarlane had produced a very good governor, but he thought it could be very much simplified. He mentioned a governor that he had taken to pieces—and he (Mr Russell) thought he knew that governor—and found it to consist of 64 parts. He himself had set about making a more simple governor, which worked horizontally, with the balls outside; and he had met the same difficulty as Mr Macfarlane referred to at page 7 of the paper— that the sensitiveness was destroyed by the friction of the stuffing-box. This he had got over very satisfactorily by taking away the stuffing-box altogether, and giving it a good long bearing, so fitted that any escape of steam was almost invisible. He did not know that it was a very good plan to have the governor balls revolving among the steam. He thought it was better to have the revolving parts outside, where they could be seen. He was not aware whether Mr Macfarlane stated the number of revolutions, per minute, that his governor ran; but that might be added to the paper.

Mr MACFARLANE said that the number of revolutions varied according to the size of the governor. For a six inch governor the speed is about 200 revolutions per minute, and for a one inch governor about 900 revolutions per minute.

Mr W. RENNY WATSON said that as one interested in such matters, he could bear testimony that the governor brought before the meeting had worked very satisfactorily, and had indeed answered all the expectations that Mr Macfarlane had claimed for it. He did not quite see the difficulty that Mr Russell had suggested—that the working parts were inside. He thought they were well in out of the way, and out of the dirt; for thus the governor needed less cleaning than usual. The ordinary governors needed frequent cleaning, running as they did at great speeds. From the experience of the governor he had had, he had much pleasure in congratulating Mr Macfarlane on the present occasion.

Mr AIRD said the Acme Machine Company had one of Mr Macfarlane's governors working for six or eight months past, and the only fault he had with it was that it was too sensitive. It worked admirably, unless when the fireman neglected his duty and let down the steam, and he found that when that took place, the engine did not only slow down, but stopped entirely, and so called attention to the fact that the fireman had neglected his duty. They worked at 60 lbs. pressure, and when the steam went down below 35 lbs. pressure, the engine stopped. It was also most useful in case of a breakdown. They had a breakdown that morning of a pulley on the driving shaft over the engine, which in falling knocked off the governor belt, and instead of the engine running away, it stopped entirely. He thought that fact should recommend the governor to any one who wished to use it.

3

Mr M'WHIRTER said his firm had also one of Mr Macfarlane's governors put on their engine, and he might say that nowhere was a good governor required more than in electric lighting, as they all know the danger of an engine running away with such machinery. For example where 100 incandescent lamps might be in use, the belt going off, it was only a matter of two or three minutes to destroy all the lamps, and do damage amounting to £20. They had had a breakdown, and the governor pulled the engine up at once. That afternoon, on a small engine of about 6-horse power, they carried out an experiment with a full load on the engine, which was at once thrown off, and the extreme variation was only 6 per cent.; and in another experiment it did not exceed 3 per cent.

Mr ROBERT MILLER said he could also bear witness to the excellence of the governor. He thought it a very good thing to have the working parts enclosed so that they were not liable to get clogged. His firm had had a great deal of trouble with small governors, and they had tried one of Mr Macfarlane's and found it to work admirably.

Mr PHILP asked what loss of pressure was due to the working parts of the governor being among the steam. He would imagine that there must be a considerable loss on this account. Had they ever tried a pressure gauge on the boiler side of the governor, and another on the engine side of it, and noted the difference when the engine was running at full speed?

Mr MACFARLANE did not think there could be any loss due to the working parts being in the steam provided; there was sufficient passage for the steam past these parts. The pressure on the side next the engine must necessarily be less than next the boiler, or there would be no use for a throttle valve at all.

Mr PHILP said that was what he would imagine. When the engine was working at its normal full speed they wanted to get the full power of the steam from the boiler; and he thought it would be retarded less by the governor were its parts outside rather than inside.

Mr MACFARLANE replied that no doubt a certain amount of work is lost by the friction of the steam passing through the parts of the throttle valve, but to what extent it is scarcely practicable to discover. In this respect, however, it does not materially differ from other throttle valves of the same class.

The CHAIRMAN then remarked that the paper brought before the Institution, by Mr Macfarlane, was a very interesting one, and it was certainly always desirable to have anything new brought up for discussion at their meetings.

On the motion of the Chairman, a vote of thanks was awarded Mr Macfarlane for his paper.

On the Construction and Laying of a Malleable Iron Water Main for the Spring Valley Water Works, San Francisco.

By Mr ROBERT S. MOORE, Superintendent of the Risdon Iron and Locomotive Works, San Francisco.

Communicated by Mr RALPH MOORE, C.E.

(SEE PLATE III.)

Received 12th, and Read 26th October, 1886.

ON the 29th April, 1884, a contract was entered into between the Spring Valley Water Works of San Francisco, California, and the Risdon Iron and Locomotive Works Company of the same place, for the construction and laying down of about 28 miles of a main water pipe to convey water from Crystal Springs Lake to San Francisco. As this pipe was made entirely of malleable iron and of larger diameter than any previously made in the country, a general description of the pipe and the construction may be of interest.

The first part of the pipe was 44 inches diameter. It stretched from the Springs to a reservoir, holding 33,000,000 galls., a mile and a half outside the city, and 200 feet above it. After leaving the reservoir the pipe was reduced from 44 inches diameter to 37 inches diameter, and again in the city from 33 inches to 30 inches.

In order to complete the work in the required time a complete hydraulic plant was designed by and erected by the Resdon Iron Co specially for this work.

The pipes were made in lengths of about 28 feet each. The plates used in the construction were of wrought-iron. They were brought from Pennsylvania in closed box cars, and were contracted for and manufactured in accordance with the following specification :—

No. of Iron.	Size of Sheet Inches.	Weight of Sheet in lbs.
7	44 × 142	327·75
7	44 × 142	325·42
6	44 × 143	354·84
6	44 × 142	352·33

Each plate to be sheared perfectly true to the dimensions given, the sides and ends being straight and rectangular with each other. The weight of each plate not to fall below 7·4 lbs. per square foot for No. 7, and not less than 8 lbs. per square foot for No. 6, each plate being perfectly flat, even, and free from warping, buckling, splits, flaws, rust, and all other defects. The tensile strength of the iron not to be less than 50,000 pounds per square inch, and the elastic limit to be not less than 40 per cent. of the above. The iron to be of American manufacture, close grained, tough and thoroughly pliable, allowing cold scarfing to a fine edge without splitting or cracking, and not to crack between rivet holes, nor between the holes and the edge of the plate while being rolled. All plates exhibiting a hard and brittle character, and that do not in every way meet the above requirements, to be rejected.

The rivets were furnished by the mills in San Francisco and were of the best burden iron, tough and pliable, that could be driven cold by hydraulic pressure without splitting or flying. They were of the following dimensions :—

Rivet where for.	No. of Iron.	Size of Body Inches.
Seam, -	7	$\frac{7}{16} \times 1\frac{1}{16}$
Lap, -	7	$\frac{7}{16} \times 1\frac{1}{4}$
Seam, -	6	$\frac{7}{16} \times 1\frac{1}{8}$
Lap, -	6	$\frac{7}{16} \times 1\frac{5}{16}$

The pipes were all made under cover, and during the progress of manufacture all the plates and rivets were protected from rain and fog. They were all made of large and small courses, being 44 × 143 inches, trimmed to the exact size. The sheets for the small courses were 44 inches by 142 inches and trimmed. When punched and riveted they formed a tight fit into the large courses. The rivet holes were punched according to the dimensions marked on the following table :—

No. of Iron.	Distance centre to centre in each row of straight seam in inches.	Distance centre to centre between two rows of straight seam in inches.
6 and 7	1·9	1

Distance centre to centre round seam in inches.	Distance centre of seam to edge of sheets in inches.	Diameter of rivets in inches.	Diameter of rivet holes in inches.
1·50	$\frac{7}{8}$	$\frac{7}{16}$	$\frac{15}{32}$

The process of manufacture of the pipe was as follows :—

The plates were deposited by trams alongside the first multiple punch where the round seam rivet holes were punched. From this machine the plates were slid to the second multiple punch, which punched the straight seam holes. These multiple punches were arranged with a travelling table between two punch heads, swung on pivots, and operated at each revolution of the machine by cams. To the bottom of the table were fitted steel racks accurately cut to the pitch of the holes to be punched. Motion was given to the table by a pawl operating on the steel racks with a length of stroke adjusted a little in excess of the pitch required by the rivets, and the table was allowed to follow back the motion of the pawl, until it landed upon another fixed pawl which determined the fixed position of the

table at the moment the plate was punched, thus giving perfect
accuracy to the holes. These machines ran about 50 revolutions
per minute and punched five holes to each head, or ten holes to both
at each revolution of the machine. The edges of the plates were
laid against stops on the table, and a bar hinged at one end of the
table acted as a clamp to keep the plates in position. By means of
the two punching machines the sheets were finished, being completely
punched ready for scarfing and rolling with little additional handling,
with the advantage of easy adjustment for all sizes of sheets and for
the small and large courses. After leaving the punches the plates
were scarfed, cold, at the corners and then swung into a set of
upright rolls with a hydraulic crane. These rolls run continuously
the size of sheet permitting of its being sprung out without stopping
or raising either roll. Each plate was rolled to a perfect cylinder
of the required diameter, and the courses pressed into each other
under a pressure of four tons. The work of riveting was done with
twelve hydraulic riveters, the pipe being swung in a vertical position
and raised or lowered by a hydraulic hoist attached to each riveter,
and controlled by the operator   The riveting was all done cold, the
rivets being placed in the holes from the outside. Where at the
end of each course the lap fell between two thicknesses of iron, the
sheets were drawn to a fine edge, through which edge upon riveting
the courses together one of the rivets of the round seam was driven
to insure absolute tightness.

The greater portion of the pipe was manufactured in lengths of
eight courses each, so that each length had large and small courses
on the ends, both ends of each length having the regular round
seam rivet holes. All round seams were split and caulked, and
straight seams chipped and caulked in first-class boiler-work fashion,
particular care being taken at the laps and for three inches along
the seams adjoining. Occasionally a length of pipe was tested at
the works under a pressure of 125 lbs. per square inch. After
being carefully chipped and caulked the pipe was moved to the
dipping yard, where it was coated with a mixture of asphaltum and
coal tar. For this purpose, two large tanks 32 feet long by four

feet deep, were constructed and placed in position side by side. They were heated by direct fire. The asphaltum in the first tank in which the pipe was dipped was kept at a good temperature, and that in the second a little less. The asphaltum in the first tank contained a larger proportion of coal tar and was more fluid than that in the second, and was used principally to insure a perfect coating and heating of the iron. The pipe was allowed to remain in the first tank 25 minutes, or until the iron was at the same heat as the asphaltum; it was then raised and hung upright and allowed to drain all the superfluous asphaltum off its surface and to cool enough to harden the coating, after which it was lowered rapidly into the second tank and withdrawn when the iron was again the heat of the tank. Thus the pipe had two coats, the first covering and connecting with the iron, the second rapidly amalgamating with the first. After again hoisting and well draining the pipe, it was again dipped and quickly withdrawn, thus joining the coat of asphalting to the second and third without heating the pipe, but only mélting the surface of the second coat, obtaining thereby a perfect coating and of greater thickness than could be obtained by one immersion. The thickness was $\frac{1}{3\frac{1}{2}}$ of an inch at the thinnest places.

The pipes were transported to the railway station nearest the pipe line on flat cars. A large quantity, however, had to be hauled for ten miles or more on waggons. The latter were arranged to carry three pipes two abreast, and one on the top; a second light waggon went behind carrying one or two sections. When the teams reached the base of a hill the second was left behind, and the horses assisted in taking over the first, and then went back for the second. By this means more work with fewer men and teams could be got done than by any other arrangement. On arriving at the trench where the pipe was to be laid, the pipes were rolled off the waggon to the ground adjacent, where they were left until ready for laying. At one point the pipe was carried over a marshy arm of the bay and over which it was carried on a bridge of trestle work a few feet above the water and mud. In this case the pipes had to be moved in lengths along the trestle work from one end.

The trench was dug by contract by white labourers. It was six feet deep by four feet eight inches wide at the bottom, and five feet four inches at the top, and was laid off accurately by surveyors.

In the construction of the work three large hills were tunnelled, the character of the ground in one of them being such that it was unnecessary to line the channel with wrought-iron piping. A lining of brick in Portland cement was substituted. This tunnel was 2200 feet long, and the wrought-iron pipe was connected to it at either end by a wall of concrete.

The two other tunnels, 1200 feet and 300 feet long respectively, were lined with the wrought-iron pipe, which was laid in single large and small courses. The rivet holes in the end of the round seam of the large course having a thread cut in them to receive $\frac{9}{16}$ inch tap bolts, while corresponding holes in the small courses were punched $\frac{5}{8}$ inch diameter and left smooth. The small course was then entered into the large one until the bolt holes were exactly opposite, and the joints made with $\frac{9}{16}$ inch bolts with hemp lappings. The bolts being screwed at the ends of the pipe through the inner and outer sheets firmly so as to bring the iron closely together. After this the end of the inner sheet was split and caulked and the laps for a distance of four inches chipped and caulked.

The tunnels, which were lined with piping, were driven about six feet in diameter in the clear, and, after each course of pipe was laid, it was surrounded and securely embedded in two courses of bricks and a mixture of brick and concrete.

Where the trench permitted the pipe was connected by inserting the small course of one length into the large course of the other and riveting together without straps; the straps being used only where the curvature of the line made them necessary. In this case the lengths of the pipe were placed in the trench so that the ends butted together or nearly so, while in the case of bends in the pipe the distance apart was from one to four inches. Before the straps were fitted and riveted to the ends of the pipe the latter were carefully scraped and entirely cleaned of the asphaltum coating for a distance of three inches at each end both inside and out, so that the strap joints made a perfect union of iron to iron. The seams of these strap joints were riveted with hot rivets, forming a good substantial head both inside and out, of a shape and proportion similar to that in the other seams, care being taken that a rivet was placed through the scarfed edges of the sheet at all the laps, for which laps the lap rivets were used as specified for the pipe. Where the curvature of the line was so great that the above strap joints were insufficient to make the pipe follow the curves, the same was accomplished by inserting one or more single large courses or such single courses intermingled with the lengths of pipe. Owing to the nature of the pipe line many hundreds of feet were laid with these single courses.

On reaching the city limits, much of the 37, 33, and 30 inch pipe, owing to the line passing through made ground, had to be connected with lead joints. In order to make the connection each length of pipe was made with larger courses on the ends, the end plates inside and outside being scarfed at the ends. A six-inch nipple or sleeve of $\frac{3}{16}$ inch iron was inserted for a distance of three inches into one of the larger end courses, and riveted to it with twelve $\frac{1}{2}$ inch rivets; the pipes were then brought together in the trench, and the end with the sleeve above referred to was inserted into the large end course of the new pipe, a lap-welded iron ring of five inches by $\frac{3}{8}$ inch iron being provided of sufficient diameter to admit of $\frac{3}{8}$ inch clear space between the inside of the band and the outside of the pipe. This space was filled up with hot lead, which, after cooling, was properly caulked. (See Fig. 1, Plate III.)

These joints have been used to great advantage in mining districts in California, where the pipe ditch is through mountainous countries, and in an irrigating scheme in the Sandwich Islands, where 7000 feet of 42 inch pipes were used to cross 36 distinct gulches, varying in depth from 200 to 450 feet. This has been running about seven years, and in no case has failed; but the above are the largest that have been laid and connected in this manner. This lead joint was patented by Mr Joseph Moore some years ago, and though used extensively on this coast, we have no instance of its giving out when properly made.

This part of the pipe stands a pressure of 75 lbs. on the square inch, and while subsidence of the ground has already taken place, the pipe has accommodated itself to it, and the joints are as tight as when first laid.

The fittings, such as man-holes, air-valves, and blow-offs, were made according to the accompanying drawings (see Figs. 2, 3, and 4, Plate III.), the man-holes being placed in some instances 500 feet apart, but usually as far as 1000 feet. Each of the above fittings were fitted with a wrought-iron ring, $\frac{5}{16} \times 3$ inches for the inside of the pipe, hot rivets passing through it, the iron of the pipe, and the flange of the respective attachment. Only hot rivets were used for this purpose, and the edge of the plate was chipped and caulked against the inside face of the casting; the rivet joints, as well as the apparatus so attached, being perfectly water-tight.

A quantity of asphaltum was kept hot along the line, and where-ever pipe-iron was exposed, due to transportation or handling, it was re-coated.

The making and laying of the pipe was stopped during the winter months. With everything in good working order, as high as 950 feet was made per working day of 10 hours; with a straight trench the same amount was laid. The quantity laid, however, depended entirely on the character of the trench, the distance being often as low as 200 to 300 feet per day.

In the entire length of 28 miles of pipe, after the water was turned on, no part had to be emptied for repairs or caulking; it was satisfactory throughout.

Since the completion of the above work, the Risdon Iron Works have made and laid 5 miles of 37-inch wrought-iron pipe, $\frac{3}{16}$ inch iron, for the City of Oakland. The line ran through a level country, and the entire work was completed in six weeks from the signing of the contract. Water was turned on a few weeks afterwards, and not a leak has occurred in the entire line.

In the laying of wrought-iron pipe, which is riveted end to end, the principal precaution is to lay the pipe in the ditch in its proper place, with the rivet holes in position to receive the rivets; then clear out a space around the joint for the men to work. After this is done, a quantity of earth should be thrown over the pipe, sufficient to protect it from the action of the sun or change of temperature between night and day. Gangs of men may then be sent to rivet up the joints anywhere it is most convenient. In other words, to leave the expansion and contraction that may take place to each pipe only. There has been no trouble from this cause where pipes were buried.

What I have endeavoured to show in this paper is the fact that malleable-iron pipes can be safely used under very considerable pressure; that they can be riveted together in long lengths without suffering from expansion or contraction; and also that they can be laid in yielding ground with a minimum risk of leakage. They are thus very suitable for waterworks' mains, or mains for irrigating purposes.

With regard to durability, they cannot be said to be equal to cast-iron; at the same time, the fact of the Spring Valley Water-works Company having decided to lay this large pipe of wrought-iron, shows that their experience of the first two lines of 30-inch pipe, extending over a period of 20 years, has been sufficiently satisfactory to warrant them in adopting it.

In the after discussion,

Mr JAMES M. GALE said he thought the paper they had just heard read was a very valuable and also a very interesting one. He was of opinion that the success that had attended the use of wrought-iron pipes in America would certainly force the subject upon the attention of the engineers of this country. There were already many wrought-iron pipes being made in Glasgow for ship-ment; and in countries where the roads were not good, and where cast-iron pipes could not be taken over them easily, he thought such pipes as those described in the paper would be of immense advantage —indeed it was already known that in such places they were being used with advantage. He had no doubt there was a great future for wrought-iron pipes, especially large sizes, and even in this country they might come to be used yet. No doubt their period of duration, particularly when they were made so thin as they appeared by the paper to be made in America, must be short, but on the other hand their cost must be very much less than similarly strong pipes made of cast-iron. It was known that wrought-iron boilers, properly proportioned for the strain to be put upon them, rarely failed within 20 or 30 years, unless by mismanagement. Now a cast-iron pipe was a very peculiar structure, and when it gave way one could very rarely trace the cause of the disaster; but a wrought-iron pipe properly proportioned gave absolute security, and as stated in the paper they could lay 28 miles of it without a failure. No civil engineer in this country would undertake to lay a pipe of cast-iron of the same length without a large number of failures; so that he thought there was an absolute security as well as cheapness in their use, for at the end of say 20 years they must have saved a good sum of money to those employing them, the interest of which might enable them to replace them. If wrought-iron pipes lasted 20 years then they must be held to have done well, and earned money for their replacement. He would like more particulars as to the substance with which the pipes were coated—asphaltum and a preparation of coal tar—as to where it came from. The dipping was graphically described, and should answer the purpose well. This general subject had recently been brought under the notice of some engineers in Glasgow by a gentleman from London, who made pipes of steel, which no doubt would be thinner than those of wrought-iron, as described in the paper. That gentleman preferred not to rivet them, but to lay them in the ordinary way with lead. He (Mr Gale) thought that the riveted joint—unless the riveting threw some strain upon the pipe, which he did not see that it would do—would be the best, especially for large pipes. He was not prepared to say more on the subject until the paper was printed, when some other gentleman might have thought over the matter, and be able to give some more information upon it.

Mr RALPH MOORE said that Mr Joseph Moore, the father of the writer of the paper, was in London at present, and if the discussion were adjourned till next meeting he would be glad to come here and give any further information required. Mr Moore had designed many of the arrangements for the work.

Mr A. MECHAN said he had listened with much interest to Mr Moore's paper, and was gratified to hear what was going on in America in this matter. In this country pipes were being made thinner than used to be the case, especially since steel had come into use, as steel pipes could stand so much more pressure than iron. He was anxious to know the amount of pressure the pipes described bore.

Mr MOORE—It was stated in the paper as 75 lbs., but it is easily calculated. The tensile strain for pipes in street is 6000 lbs.; for work where a leak would not damage property, the tensile strain is from 10,000 to 12,000 lbs.

Mr MECHAN said he might mention that his firm had made pipes of steel 12 inches in diameter, and only one-eighth thick, with double lap riveted longitudinal seams $1\frac{1}{2}$ inch pitch and $\frac{1}{4}$ rivets, and when put under a hydraulic test, withstood a pressure of 760 lbs. The machine failed to burst it. The firm would have liked to carry the test further, but were afraid to break the machine. Since then he had been making inquiry, and had been told that the pipe being under compression at the two ends was the cause of its standing so much pressure. The gentleman who told him so could not explain why that was, but said that if it had been tried like a boiler with ends closed it would not have stood the pressure it did. That he could not understand. However, the fact was this—steel stood a so much higher tensile strain than iron, that steel pipes could be made much thinner than iron pipes. He had observed recently that Mr Hamilton Smith read a paper in London where pipes were stated to have been made only one-sixteenth thick, and only dipped in asphaltum. He would like if Mr Moore could tell them what composition the asphaltum was made of. The great difficulty with such pipes as those described in the paper was to know how long they would last when covered up under the soil, for that would settle the question as to whether or not they would come of use. He did not think they would compete with cast-iron pipes if made below a certain size. They might do for mains or for export, but he did not think that small pipes of wrought-iron could be made to compete with those of cast-iron.

Mr GALE asked Mr Mechan what was the price charged per ton of the pipes he had referred to.

Mr MECHAN did not know that he should give that information at present. It varied with the thickness and diameter of the pipe.

Mr HENRY M. NAPIER said he had the pleasure of seeing the pipes described in the paper while in San Francisco last Spring, and also the plant specially designed to do the work. What took his fancy was, first of all, the beautiful plates—and he thought those who rolled plates in this country might take a lesson from them. There was no buckling in any of them. Again he was struck with the very exact work of both the punching and rolling machines, by which the plates seemed to fit the one to the other without the least trouble; and he also wondered why, if wrought-iron pipes suited on the other side of the water, they had not ere this been made in Glasgow or Scotland. He could not help thinking but that it was the necessity of the case there; that it was, in fact, a development—Mr Darwin's theory put into pipes. They really could not carry cast-iron pipes all the distance required there; but wrought-iron they could. Then it was most remarkable to see those long pipes, one piece added to another until they were 28 feet in length, and these carried on mules' backs. Defective riveting could not have stood such rough usage. Now with regard to the protection of these pipes. In shipbuilding they were at very great pains to scrape the scale off steel plates before painting them; but he asked was it not possible for the steel makers to make the scale a little harder, or of such a nature that it would stick on? Was it not possible to submit the plates to a high steam pressure whereby they might be coated with magnetic oxide and thus require no scraping and but little painting?

Mr RALPH MOORE said the asphaltum was got within 40 or 50 miles of San Francisco, and was a natural asphalte.

The discussion was then continued till next meeting.

The discussion on this paper was resumed on the 23rd November, 1886, when Mr J. M. GALE introduced

Mr D. J. RUSSELL DUNCAN, London, who said that, with the kind permission of the Institution, he would offer a few observations. The fall to be dealt with was given as 200 feet, and the diameter of the pipes under this head 44 inches, length 1½ mile, stretching from the springs to a reservoir. From the areas and weights tabulated he calculated the mean thickness of the plates to be ·1875 and ·203 inch, the mean weight per square foot 7·5 and 8·12 lbs.; 7·4 and 8·0 lbs. were given as the minimum weights per square foot; it was therefore apparent that the margin of weight and thickness allowed for unequal rolling was 1½ per cent. under the mean dimensions, and it might be supposed that the writer of the paper allowed a like margin as the maximum over the mean limits. The margin he proposed to allow on similar plates was 3 per cent. under and over the mean. As the resistance of pipes to internal pressure should be calculated on the minimum thickness of the plate, the figures in the cases under consideration became

$$·1875 \times ·985 = ·1847 \text{ inch}, \quad \text{and} \quad ·203 \times ·985 = ·2 \text{ inch}.$$

The diameters in the calculation for resistance to internal pressure should be the maximum at the widest part of each course of plates,

A = nominal diameter of pipe,
B = maximum diameter for calculation,
= 44·375   and   44·406.

Applying these thicknesses to the usual formula, and making a deduction of 40 per cent. for loss of strength in the solid plate due to riveting—

T = thickness in inches,　$T = \dfrac{\cdot 433\ H \times D \times F}{\cdot 6 \times S \times 2} = \cdot 1847''$ and $\cdot 2''$

H = head in feet,　　　$H = \dfrac{T \times \cdot 6 \times S \times 2}{\cdot 433 \times D \times F} = 192'$ and $208'$

S = tenacity in lbs.　　$S = 50,000$ lbs.

D = diameter in inches,　$D =$　　　　　　　$44 \cdot 375''$ and $44 \cdot 406''$

F = 3 (in America) factor of safety.

Working according to the English practice for mild steel, we should use a factor of safety = 5, and therefore the values of H would become 115 feet and 125 feet, instead of 192 feet and 208 feet, showing that in England they should work such pipes at lower pressures than in America. Again, taking the head as given in the paper, and applying the same formula, but in this case allowing for the strongest possible riveted joint—viz., 70 per cent. of the strength of the solid plate—the value for T would become

$$T = \frac{\cdot 433\ H \times D \times F}{S \times \cdot 7 \times 2} = \frac{\cdot 433 \times 200 \times 44 \cdot 5 \times 5}{50,000 \times \cdot 7 \times 2} = \cdot 2785 \text{ inch.}$$

The maximum external diameter, 44·5 inches, is assumed to allow for twice the thickness of the plate to be calculated. But, as ·2785 inch is the minimum thickness of plate, the maker's margin of 3 per cent. for unequal rolling must be allowed, therefore the mean thickness of plate for the specification would become $\dfrac{\cdot 2785}{\cdot 97}$ = ·287 inch, and the weight in lbs. per square foot would be 11·48, or 43½ per cent. thicker and heavier than the plate specified in Mr Moore's paper. It would thus be seen that the English practice for wrought-iron pipes was one which gave greater strength and weight than was customary in America. The thickness, ·287 inch, already referred to, was for wrought-iron; if mild steel plates were used for such a pipe, the head of 200 feet remaining the same, the mean thickness would be ·211, the tenacity of the plates being

67,200 lbs. per square inch, elongation 20 per cent. on 8-inch test
bars. This came within 4 per cent. of the heaviest plate specified
by Mr Moore, and might be said to be practically the same, therefore the difference between the American and English practice
showed that in the case under consideration (internal diameter and
working head remaining the same), a pipe which would in America
be made of wrought-iron on American rules, would in England be
made of steel of the same thickness on English rules. With regard
to the riveted joints, according to the rules to which he worked, the
pitch and lap differed slightly from the corresponding details in
Mr Moore's paper. The differences were shown in the following
table :—

	Diameter of Rivet.	Pitch.		Lap.	
		Circular Seam, Single Riveted.	Longitudinal Seam, Double Riveted.	Circular Seam.	Longitudinal Seam.
A	$\frac{7}{16}$	1·5	1·9	1·75	2·75
B	$\frac{7}{16}$	1·25	2·0	1·5	2·75

A line shows the figures taken from the paper.

B   „   „   according to his system.

With regard to the caulking of the seams, it was said that the edges
of the plates were split and caulked. He understood from this that
a kind of V was formed on the edge of the plate by means of a
sharp pointed tool, and afterwards stubbed or fullered down by
another tool. He had been shown a method of bevelling the edges
of such plates which presented a good surface for caulking with an
ordinary tool, and was applicable to thin plates. If this system
were adopted in both circular and longitudinal seams, no chipping
of the edges of the plates was necessary. The writer of the
paper did not give any particulars as to how the tests were conducted; this was somewhat important, as pipes in long lengths of

28 feet could not be conveniently tested on the machines usually employed for cast-iron pipes. With regard to the test pressure of 120 lbs. per square inch, this was equivalent to only $1\frac{1}{2}$ times the working pressure. He allowed a test pressure of not less than 1·75 times the working pressure, which was the test laid down in the rules of the Manchester Steam Users' Association for testing steam boilers, the work on which was very nearly identical with that on riveted water mains. The next point he desired to touch upon was that of coating the pipes. He understood that two tanks were employed, and he gathered from the paper that the pipes were completely immersed in the liquid composition. He was greatly indebted to Dr Wallace of this city for having made some experiments upon compositions to be used in a bath which he had designed for the purpose; and without going into the details of the apparatus, he might mention that by using one bath only, upon a special plan arranged on a principle of alternate dipping and cooling by rotation of the pipes during partial immersion, pipes could be coated to a suitable thickness, and in less time than the writer of the paper required. Some steel pipes, 24 inches internal diameter, about 25 feet long, were recently inspected at Bury St. Edmunds, part of a lot now being supplied to the Corporation of that town, which were constructed in Glasgow to his specification ; and he understood from an American engineer, who had inspected the 44 inch pipes at the Risdon ironworks, that the 24 inch steel pipes which he saw at Bury St. Edmunds were quite as well coated as those made at San Francisco. He referred to this to show that the simpler and less expensive system now working in Glasgow was in every way equal to the American practice.

Mr Joseph Moore replied that in California they had to put up with what they had and did not go into great niceties. The course they had adopted was to feel their way along, and the paper gave the conclusion at which they had arrived. In this case the pipe was made to bear 6000 lbs. tensile strain on the rivet-seam. From what they had running they made up their minds that this was a safe pressure for anything in the shape of waterworks. When

first they began to make these pipes the tensile strain was as much
as 18,000 lbs. per square inch.   The pipes were light and easily
handled, and if they did burst they were easily repaired.   Of
course, at that time in California they did not care much whether
a break took place or not, as the water did no harm.   Gradually
they reduced the tensile strain to 14,000, then to 12,000.   The
Spring Valley Water Co. had had a long experience with these pipes,
There were two lines of pipes, over 20 miles each, standing a strain of
12,000 lbs.   The pipes referred to in the present paper were only
strained to 6000 lbs. per sq. in., and this is perhaps a safe strain for
water mains.   It would be seen that the factor of safety was now three
times that employed when pipes of this description were first used.
In riveting they had to use what tools they had and these had
answered fairly well.   Only they were afraid sometimes to go to too
heavy pressures on account of the caulking of the seams.   The
rivets were never closer than mentioned, and they had a 1940-feet
pipe 12 inches in diameter, running six or seven years, wider riveted
than the rest without having any trouble with it.   In point of fact
they never were very scientific but they never had any failure.
In regard to testing, they tested one or two pipes until the
seams blew water to an extreme height, and they concluded
after that that there was no danger at all under the pressure they
put upon them.   The pressure of 125 lbs., they must understand,
was put on not to find the strength of the pipe but to find
imperfections in the riveting or caulking.   It was entirely on that
account that they tested the pipes; but the fact that they finished
the pipes and turned on the water and that no part of them had
been uncovered, showed that they were not wrong in their conclusions.   The coating which they gave the pipes consisted of
asphaltum, of which there were lakes in that region, and it was
found to do admirably.   In fact he had taken a pipe after it had
been many years in the ground, reduced it from 19 to 12 inches,
and made a new pipe of it.   Only great care must be taken that
there was no rust on the pipes, and special precautions were adopted
to insure that this was the case.   With the view of testing this

action, a pipe of 36 inches diameter, which had been running for a number of years, was examined, to note the action of the water inside. It was found that on the bottom side the scour had taken off the coating, and more or less worn down the rivets, but in no case had the pipe rusted as it would have done on the outside. The coating was of a tough consistency and would not melt under the heat of the sun. It would keep he did not know how long, but they knew from past experience that it was perfectly safe for 20 years.

Mr J. M. GALE asked Mr Moore if he thought the preparation of coal tar used in this country for similar purposes would be as efficient as the asphaltum used in California ?

Mr MOORE could not tell, nor could he describe what the asphaltum really was. It was quite soft, like pitch, but not brittle, it did not tear. On being retorted in a cast-iron vessel it gave off coal oil of all kinds—kerosene, lubricating oil, and coal tar ; indeed he did not know what it did not give off, and the residue was tar. There was a large quantity of sand in it. It had to be heated enough to drive off the lighter oils. If it was not boiled a great deal it remained perfectly soft. It would be an easy matter to get a sample of it.

Mr GALE thought the success of the pipes must depend on the coating—the length of time they remained in the ground must depend on the protection given to them by it. As far as duration went in this country the coating did not last above twelve years. Even before the expiry of ten or twelve years here and there a carbuncle or tubercle appeared on them. He had an opportunity recently of looking into some pipes laid for the Loch Katrine Water-works more than 26 years ago, and in some cases the coating had almost failed to protect the pipes. The coating had not disappeared, but the oxide seemed to exude through flaws in the coating, and then to coalesce until they formed a mass of tubercles. Anything of that kind happening in these wrought-iron pipes would oxidise into holes in them very soon. The coating at San Francisco must be very different from anything they had any experience of in this country. He was very much astonished at a statement he noticed recently in regard to steel sleepers. It was stated that from the experience with these sleepers in this country they would last 20 or 25 years. He did not know whether these sleepers had been protected in any way; but, perhaps, some of their engineering friends in that line would be able to enlighten them on the subject. Doubtless these were as thin as the pipes of which they had been hearing—(A voice—$\frac{3}{16}$ths), so that the pipes would last for a like period, although he would not have expected it, and this opened up an important matter in regard to thin sheet-iron. In another place he saw it stated in regard to steel sleepers that from observations made on the Continent they would last the lives of three or four wooden ones—that was 30 or 40 years. It occurred to him that this lengthened life might be due to the vibration under which they were kept by the passing of heavy loads, as it was certain that railway iron did not rust so quickly as iron exposed to the atmosphere without any such vibration. He had no doubt, however, some of the railway engineers would be able to give them information on the subject.

Mr JOSEPH MOORE explained that it was found that the coating did wear off, but that the iron did not rust inside these pipes.

Mr W. RENNY WATSON congratulated Mr Moore upon the great advance made in the production of these pipes, which were a much better article than the one he saw when he visited California eight or nine years ago. The common-sense rules which had been adopted had been proved to work very well, as was shown by the results which had been put before them. It was hardly fair, he thought, to consider water pipes in the same way as steam boilers, for the one was not subject to the same great variations of heat and cold as the other, and didn't, therefore, require the same factor of safety. As to the remark Mr Duncan made in regard to ring-seam joints, he might say that he saw at the Bloomfield Mines in Nevada a 22-inch pipe, working with a head of water of 324 feet. It was made of $\frac{3}{16}$th plates, like a common boiler, and had no rivet at all in the ring seams, and yet it stood all that pressure. One would have thought this pipe must have burst, but it didn't.   In answer to some inquiries he was told that it had leaked a little at first, but that several cart loads of meal and sawdust were passed along with the water and these having found their way into all the little interstices made it perfectly tight and most suitable for mining purposes, for which it was lasting extremely well.   In that connection he might mention that near Lake Michigan wooden pipes were used, four sizes being got, core after core, out of a single 30-inch square log of wood 12 feet long.   It looked expensive, but then they had the wood for nothing.   The pipes, which were coated with asphalt, were used with remarkable success both for gas and water in several comparatively large cities in that neighbourhood.   Perhaps some gentleman present could tell them something about them.

Mr GALE believed they might assume that wooden pipes were the first pipes made.   On some alterations being made recently at Regent Circus, London, pipes, laid by the New River Company, made of elm, were exposed after having been in the ground for about two hundred years.   Wooden pipes, being cheap, were generally used in the United States as a beginning; but they did not last long, and were always replaced with cast-iron ones, as the towns increased in importance.

Mr HADDIN presumed that the pressure on the pipes would be constant, and this he thought would be in their favour, as cast-iron pipes deteriorated in consequence of the varying pressures, and concussions of heavy overhead traffic.   The constant flow might also account for the cleanness of the pipes, the gravel or sand as it went along scouring the metal.

It was agreed that owing to the importance of the subject the discussion should be adjourned till next general meeting.

The discussion on this paper was resumed on the 21st December, 1886.

Mr JOHN THOMSON said as he had moved the adjournment of the discussion at last meeting, he supposed he might begin it that evening. The author of the paper under discussion seemed to have for a long time taken great interest in malleable-iron piping, as on looking over the Institution's Transactions he observed that the last paper brought before them by Mr Moore was in 1883, the discussion of which had also been twice adjourned. He thought it must be very gratifying to Mr Moore that his papers elicited so much discussion. Mr Moore had again given them a very interesting paper, the result of his experience of a further development of water piping in San Francisco. He had very little to add to the last discussion, but he thought the main point in the matter had not yet been touched upon at all—the question of cost. Neither Mr Moore, Mr Duncan, nor Mr Mechan, who had had some experience in the manufacture of such pipes, had given any indication of what the cost per ton, per yard, or per foot of that class of pipe was. That really was the crucial question, including, as it did, the lasting of the pipes in the ground, and their renewal, all of which entered into the question of cost. According to Mr Moore's experience these pipes seemed to be lasting very well. The mode of coating seemed to him to be the main point. By general experience it was ascertained that wrought-iron had a very great tendency to oxidise; and therefore, unless these pipes were very carefully and efficiently coated, they could not last any great length of time. No doubt wrought-iron pipes would be very convenient and useful in San Francisco, where those referred to by Mr Moore were used; but he thought that Mr Henry Napier had spoken truly when he said that it was just another development—Mr Darwin's theory put into pipes—a development due to the natural position of the place, because the conveyance of heavy cast-iron pipes from Pittsburg, or any other iron district in America, to San Francisco, would be very costly. He did not think, however, that the people there had yet gone in wholly for wrought-iron pipes, for recently there was a scheme for bringing another supply of water into

6

San Francisco by means of English, or, he rather hoped, Scotch cast-iron pipes, the idea being to take them thither by sea. The import duty in America on pipes, or indeed on any manufactured article, was very heavy, and of course militated against the introduction from abroad of this class of goods. It was a very extraordinary thing that in these wrought-iron pipes the water seemed to take the coating off the inside, as Mr Moore stated that even the rivet heads were worn away to some extent by the action of the water. The coating, as described by Mr Moore, seemed to be very efficient. They had nothing of the kind in this country, and it would be interesting—as suggested at last meeting—to have a sample of it here. It was a natural product, and Mr Moore said there were lakes of it in America, so that much of the advantage claimed for the pipes must be due to the coating. Mr Duncan had said that he could get pipes as well coated here, but he (Mr Thomson) was afraid it would be a very expensive process. Here they had to get over the coating very much quicker and cheaper than what Mr Duncan seemed to suggest, and he did not know how it would do to have to turn each pipe round on a spindle, and expose it to a cooling action of the atmosphere. He might remark that the use of wrought-iron pipes was not new. His firm had made them for special purposes, and no doubt for special conditions they were very useful; but whether they would be useful for a permanent water main was quite another question. He was not sure that he could agree with what Mr Gale had said about malleable-iron pipes if they stood for twenty years. Of course, it was evident on the face of it, that if they had earned as much interest in these twenty years as would pay for their relaying, they would require to be very cheap at first—something like half the cost of the cast-iron pipes, or less. Now, he hardly thought that wrought-iron or steel pipes could be produced at so low a price. Mr Gale, on a former occasion, had told them about the Kimberley pipes, these were not riveted; they were 14 inches in diameter and laid with collars and lead joints, and were used for conveying water to the Kimberley diamond mines, and he understood they suited the purpose quite well. Those pipes referred to by Mr Moore were all riveted, and the riveting seemed to be very well done, the apparatus employed being apparently well adapted for the work. Mr Mechan told them of some pipes he had made, and which had been tested in his firm's works ; but he must say that the test was not such as would satisfy inspectors in regard to cast-iron. There was a good deal of leakage when the water was put on, even without pressure, and there was considerable difficulty in making them sufficiently tight to get up the pressure referred to. He had put a 12 inch cast-iron pipe into the same proving machine as Mr Mechan's 12 inch steel pipe, immediately afterwards. It was half an inch in thickness, and stood 800 lbs. pressure before bursting ; indeed it did not burst until the pressure indicated on the gauge was 2200 feet, so that they did get cast-iron pipes to stand a pretty good pressure also. They had themselves tried riveted pipes, but never could get them to stand the test at the seams perfectly. They had galvanised them, but that had not any specially beneficial effect, so that he could not say that those pipes that Mr Moore or Mr Mechan brought under their notice were quite satisfactory for the requirements of engineers in this country. They might do for San Francisco, but they would not pass Mr Gale's inspection here. They knew that wrought-iron under the ground corroded very quickly, and had Mr Foulis been present, he could have told them that he found it necessary to imbed his wrought-iron service pipes in pitch before burying them in the ground. Proof that wrought-iron in some kinds of ground decayed quickly had come to his hand lately. His firm had been employed to examine and report at Edinburgh on some 2 inch tube, and they found that those pipes were completely worn out in nine months. He exhibited an example of one of the pipes which showed holes right through. It was right to state that those pipes were not coated, and when taken out were quite clear inside. They had been laid in ordinary made ground round the Edinburgh Infirmary. This was a particular experience of the effect on a wrought-iron pipe put into ordinary soil. It all came back, however, to the original question of cost, and if it should be found that these pipes can be made cheaper and last longer than cast-iron, then steel or wrought-iron pipes may be adopted. Mr Moore, in his paper read before this Institution on 24th April, 1883, said—"It is not to be supposed that the author recommends wrought-iron pipes at all times where cast-iron can be got for nearly the same price, or where a permanent work is laid down; but it is believed that, from the experience gained with it, its cheapness, and its durablity when properly constructed, and the despatch with which it can be laid, there are many cases in which its use would pay. In temporary or preliminary work, there is no doubt it is worth serious consideration." He thought Mr Moore had very fairly stated there the real position of the matter. At the mines and other preliminary works they had been found useful for the purpose intended. Altogether, he must say, that he had read Mr Moore's paper with great interest, and he thought the Institution were indebted to the Messrs Moore for these communications sent to them from the other side of the water. He would like to add that the oldest cast-iron pipe that he had had the privilege of seeing was laid at the palace of Peterhoff, at St. Petersburg. He believed it was about 80 or 100 years since this pipe was laid. The pipe referred to was about 14 or 15 inches in diameter, and was cast in short lengths of 4 or 5 feet long, and fixed with snugs and bolts. A considerable part of it was exposed to the atmosphere, being partly laid in a ditch and partly underground. It was used to carry water to the fountains, and the pressure that those pipes were continually under was very heavy. He did not know where they were cast.

Mr HOWARD BOWSER said that his remarks would be very much in the line of those of Mr Thomson, although he thought this subject should be treated more from an engineering point of view than from a commercial one. As to the latter point of view it would be seen that the question entirely depended upon the margin of safety that was allowed in cast-iron or malleable-iron. In cast-iron a very considerable margin of safety was allowed, and if the same margin of safety were allowed in malleable-iron, it would be perfectly evident that it could not contend with cast-iron. Another point mentioned by Mr Moore was the rapidity with which those malleable-iron pipes could be laid. He would take Mr Moore's authority upon that point as worth a great deal; but he had seen in Glasgow streets, seventy 18 inch pipes, 12 feet long, laid in ten hours. Now he did not know whether Mr Moore would say that was a rapid laying of pipes or not. These were bored and turned pipes, requiring no jointing. With regard to the question of riveted joints, it would be admitted that they could not be made tight without caulking. Now, what did caulking mean? It meant the splitting of the edge of the pipe, and staving it up so that the leaks between the rivets were stopped for the time being. By what were they stopped? By the caulking. It was quite clear that the caulking in the inside of the pipe must be in contact with the water, so that, instead of the thickness of the pipe, the thickness of the caulking is all there is to contend against pressure of the water or rust. Of course if they added coating to coating it might protect the surface altogether, but he remembered being in Paris in 1855, and, going into the water works yard, found a quantity of pipes lying which appeared to be about one inch thick. These were all thin malleable-iron pipes, and had been coated with asphalte until the coating upon them, both inside and outside, was one inch thick, so that one would think the coating should have been a perfect protection for the iron. Notwithstanding the pipes had failed, and many hundreds had been taken up after being a very short time in the ground, to be replaced by cast-iron pipes. These were great difficulties to overcome, and it was for mechanics to meet them; but if the day ever came when malleable-iron and steel pipes were preferred then they would require to face the necessities and requirements of the day. Certainly cast-iron was a very uncertain metal; still if they put a limit to the life of malleable-iron pipes of twenty years, and compared that with the actual life of cast-iron pipes, he was satisfied the malleable-iron pipe would be placed at a great disadvantage in comparison with cast-iron. He thought it would be easy to show that the cast-iron pipes would maintain their position against those made of malleable-iron.

Mr C. C. LINDSAY said that he had read the paper with pleasure and profit. There was only one point to which he would draw attention, and on which he wished to ask a question. It was with reference to the jointing. The joint showed an inner ring and an outer collar leaded in the usual manner. It seemed to him rather a crude practical method of jointing, and was objectionable because of the decreased area of flow and the friction to which it gave rise at each joint. He considered that there was no difficulty in making a good practical joint without an inner ring, and instanced the Kimberley Water Works pipes, to which reference had been made. In 1880 when the plans of these works were in hands, the engineer, Mr Buchanan, had consulted him as to the pipes, which had to be carried up country, and also as to the joint. He proposed two forms of joint—one with a weldless faucet riveted to the welded tube forming the body of the pipe; the other was a simple wrought-iron collar. The latter form was adopted, and had been found to answer very well as regards practical jointing, free full water area, and repairs. He would be glad if Mr Moore would give some par ticulars as to the reason for adopting the joint, and as to the life of the joint and inner ring.

Mr JAS. WATSON (Dundee), on the invitation of the Chairman, said he had come to the meeting to gain information, and he did not know anything he could say that would add to what they had already heard. He had had the opportunity and experience of working one of the most difficult lines of cast-iron mains that existed in this country. He knew that cast-iron was liable to many diseases. From the time it came from the cupola, during the process of manufacture, and often throughout its working life (especially where the ground bearing was irregular, and the pressure great), consider-able difficulties were to be contended with. He should have liked to have had some information—had he been a member of this Institu-tion—on various points. He would have asked information on three points had that been the case—two of them seemed to him of great importance, and the other not quite so important. The first point—referred to by Mr Thomson—was as to the question of the lasting of the malleable-iron pipes compared with that of cast-iron pipes; and the second was with reference to the quality and nature of the water conveyed by malleable-iron mains into San Francisco. There was no doubt that the chemical constituents of the water largely determined the nature and extent of corrosion of cast-iron pipes, and particulars or analysis of the water would on this account be useful. Coating only protected pipes for a time; he found that in a large main laid down twelve years ago, the coating was now quite worn, and the surface considerably oxidised. He knew of no coating which entirely prevented this oxidation and incrustation from going on, and, with soft water, going on rapidly. The theory of the late Dr Angus Smith, that cast-iron pipes took coating more permanently than malleable-iron, seemed to him well founded, the granular nature of cast-iron, if properly heated in the coating bath, rendered it possible to so far incorporate the coating mixture with the surfaces or skins of the casting; such coating, in the case of malleable-iron (by reason of the closeness and disposition of the metal) was nothing more than a mere skin or contact coating, liable to be removed by the slightest friction or abrasion. If malleable-iron pipes were only to last twenty years, the initial cost must be very largely under that of cast-iron. He was at present lifting two miles of cast-iron piping made in Glasgow in 1845. They were 15 inches in diameter, had been forty years working, and the iron of this line of main pipes was just as perfect as the day it was laid down. It was greatly oxidised, however, so that the pipes were largely filled with oxidised matter, which seriously diminished the delivering power of the main. The other part he would like information on was a minor (though not unimportant) one—this was as to making repairs with malleable-iron pipes *in situ*. It was well known that they could readily repair cast-iron pipes when an accident occurred. He would like to have heard how in the case of some of the rivets in Mr Moore's pipe giving way, or a fracture in the iron taking place, how repairs were executed. It did seem to him that if one had to put boilermakers into a trench to cut malleable-iron pipes in a hurry, that the method and time

were matters they would like information upon. He himself had had
something to do with the steel pipes referred to by Mr Thomson.
He thought Mr Thomson had stated the results of the experimental
tests of the steel tubes fairly enough ; as to his statement regarding
the test of the cast-iron pipe, it was very satisfactory so far as it
went, but he ventured to say no matter how carefully foundry work
was done, or the strength of the iron specified and measured, or the
finished casting proved in the testing machine, yet when subjected
to work involving strains greatly under those withstood by the
foundry tests, cast-iron often failed in the most unaccountable way.
He agreed with Mr Thomson that unless they could get malleable-iron
or steel pipes to last for a reasonable length of time, they would not
come into general use for water works purposes. He thought that
the nature of the soil had much to do with the life of a pipe in the
ground; and that malleable-iron was much more liable to deteriorate
by the action of soft water than cast-iron was. He, however, was
not without hope that steel pipes, which had many obvious advantages in point of strength, ductility, &c., might yet be made and
coated to compare not unfavourably, both as to wear and cost, with
cast-iron pipes. He was very pleased to be present to hear the
discussion that night, and to come from Dundee for that purpose.
He had to thank the meeting for allowing him to make these
remarks.

Mr JAMES M. GALE said he believed there was an equally old water
main to that at St. Petersburg, in use in Edinburgh at the present
time. The first cast-iron pipes that were laid in Edinburgh were
put down about 1780, and with the exception of being considerably
corroded, the metal seemed quite sound still. They were probably
laid in a clayey soil, not in ashes or anything that chemically would
act upon them. He thought that the pipe Mr Thomson had shown
had evidently been acted upon chemically, or the inside would have
been as bad as the outside was.

Mr THOMAS KENNEDY said that the worst corroded pipe he had
ever seen was one he scraped at Whitehaven. It was an 11 inch
pipe, made by Mr Bowser's firm, and the corrosion was entirely in-

side ; in fact it was so corroded that the local engineer thought that we would scrape the whole pipe away, yet the hard portion outside was so thin that a man actually put his pick through it; and that pipe they had had working with a water engine upon it, at 50 lbs. pressure, which showed the excellency of the metal. He believed that if that pipe had been coated it might have had a life of 15 years more. Some years ago the Assistant Surveyor at Aberdeen recorded the results of scraping a small water main on the Water Works there, by hand scraper. He found that near the junction with the trunk main the corrosion was heavy ; but as they scraped towards the dead end the corrosion lessened when there were side branches. In a considerable length of the main, where there were no branches, the corrosion was uniform, but further on, as branches were passed, the corrosion lessened. He believed that oxygen in the water was the chief agent in the corrosion of pipes, and that the more water that passed through mains, the quicker did they corrode.

Professor JAMES THOMSON said, as to the corrosion by the oxygen in water, it would appear that carbonic acid was also necessary to that corrosion. Many of the members present knew that iron would not corrode if kept in lime water, but that it kept bright ; and that was attributed to the absence of carbonic acid. For instance, the steel wire sounding lines invented by his brother, Sir William Thomson, were always kept in a tank filled with lime water. It might be a question with regard to the efficacy of Portland cement in preventing rust, whether it was the lime in it that prevented the corrosion, or whether it was merely the coating in keeping the bilge water off the angle irons and the plates—whether it was the mere coating or some effect from the presence of lime in it. This was merely a suggestion which he threw out for their consideration.

Mr JAMES M. GALE said that what seemed to contract the area of water pipes after being subjected to wear, was not so much oxide of iron as clay, or peat, or any deposit that might be in the water. The deposit has been found to consist of not more than 10 or 15 per cent. of iron rust. The great bulk of what came out of water pipes was simply mud taken up by the water.

7

Mr CHARLES C. LINDSAY remarked that in a paper delivered by Mr Jamieson some years ago before the Institution of Civil Engineers, the number of experiments and illustrations of the different forms of corrosion in pipes was extraordinary, and a perusal of that paper would be very instructive to any one seeking information on the subject.

The CHAIRMAN inquired whether any tests had been made of the iron of some of the old pipes taken up, and referred to by some of the members?

Mr KENNEDY replied that if they attempted to make a mark with a chisel on some pipes they broke off; but those old pipes had to be cut through.

The CHAIRMAN said that he could not quite see what steel sleepers had to do with steel or malleable-iron pipes. He thought that steel sleepers might very well form the subject of a separate paper. He did not promise to give such a paper himself; but nevertheless it would form a very interesting subject. With regard to the coating of malleable-iron pipes, referred to in the paper, it must be a very superior coating used in San Francisco. As a natural product it seemed to be far superior to any artificial coating made in this country. With regard to leakage. If a leak took place in a cast-iron pipe, in the body of the pipe, it was a serious matter. If only at the joint, either in 'cast-iron or malleable-iron —so far as any structure he had had experience in—if the leak was very small, say one-sixteenth of an inch, it would take up of itself. So far as his experience went—and it had been considerable of the smaller sized cast-iron pipes—a very great deal depended upon the nature of the soil, and whether they were exposed to the atmosphere; and also the purity of the air and the constituents of the water had a good deal to do with their durability. For instance, in the neighbourhood of Glasgow, about St. Rollox, they could not get pipes to last long, nor anything constructed of iron, owing to the impurity of the atmosphere. He had never known cast-iron pipes last for more than 40 years, and be serviceable, if laid in ordinary soil. He had taken out a good many pipes of from four to nine inches diameter, and the iron of those which had been in use for about 40 years was worn almost through, thinned to about one-eighth of an inch, and very much corroded. Cast or malleable-iron pipes laid through made up ground would not last long, especially if the ground had been an ordinary tip of ashes, they would not last longer than twelve months. If they laid them there they must surround them with clay to prevent them being damaged by the impurities of the soil. Mr Gale had referred at last meeting to the corrosion of iron and steel sleepers and rails, and had suggested that the vibration to which they were subjected diminished it: he could assure them that rails, whether subject to the action of passing trains or whether lying dead, with no traffic at all passing over them, corroded. The action of the trains simply shook off the scale and prevented it from accumulating, Of course the corrosion was greater in some places than in others. At St. Rollox, for instance, he had taken-up rails that had been in the main line about 10 years, and the loss from corrosion was about one-sixteenth of an inch on each side, apart from the wear at the head of the rail; he found that by cutting the rails into sections and planing them. About Beattock, however, where the air was pure, there would be little of that corrosion. Then with regard to steel sleepers, he had had them so short a time in use that there was really no opinion to give about them as regards rust, but in half-a-dozen years there will be some reliable data to be got as to the wear from corrosion of steel in that position.

Mr W. J. MILLAR (the Secretary) said that in connection with the reference to peaty water, he might say that he understood the boilers of the steamers on Loch Lomond remained largely unaffected by corrosion even after long periods of use. A kind of protective coating being formed upon them, which some regarded as due to the peaty nature of the water used for feeding with, combined with the grease or oil which found its way into the boiler with the feed water.

The CHAIRMAN said that locomotive engineers preferred peaty water to hard water to make steam with. Where they had to use a very hard water the tubes of their boilers leaked in a very short time; but if they had peaty water they had no trouble in that way, so that peaty water was preferred for locomotive purposes.

Mr J. M. GALE said that peaty water was usually soft, and hard water would coat the inside of the boilers. When he referred to peat as a substance which helped to contract the area of pipes, he did not refer to it being mixed with grease in hot water. He meant that peat sometimes formed the main part of that substance of which those tubercles found on water pipes was composed; but these entirely depended upon the character of the water passing through the pipes. He had seen in Ayrshire, pipes 30 years in use, and the corrosion about an eighth of an inch thick only.

Mr BOWSER said that Mr Gale would remember the Cranstonhill water pipe. When it was lifted it was thickly coated with deposit, but the iron was as good as the day it was laid. The thickness of the pipe was equal to the work required of it in Glasgow.

Mr GALE said that pipe was laid in 1809, and only lifted in 1860. It had to be lifted because it was flange jointed, and these joints could not be kept tight.

Mr T. D. WEIR said he had had occasion a few years ago to lift 300 or 400 yards of water piping at Montrose, which had originally been 12 inches in diameter. The part lifted was close to the reservoir, and he found that it had diminished in area from 12 inches to 10 inches, since 1857, when it was laid. The pipe had not been coated at all.

Mr RALPH MOORE said he could not tell them how much he felt gratified by the kindly and candid criticism of the paper. He arranged at last meeting that the Secretary would send to his brother, Mr Joseph Moore, the report of that night's discussion, and he would make some remarks thereon, which could be produced afterwards. It would be out of place for him to say anything further about the paper. His own experience with malleable-iron pipes was only as seen in California; and those seemed to him somewhat more serviceable than they appeared to Mr Thomson. If they could be made so cheaply as to enable them to be replaced in 20 years, they might compete with cast-iron pipes. He had no doubt his brother would fully explain all the points of information required. He again thanked them for their kindly criticism of the paper.

On the motion of the Chairman, a vote of thanks was passed to Mr Moore for the valuable and interesting paper he had brought before the Institution.

———

REPLY by Mr JOSEPH MOORE on the Discussion of Mr Robert S. Moore's Paper on the " The Construction and Laying of a Malleable Iron Water Main for the Spring Valley Water Works, San Francisco."

This paper of his son might be looked upon as supplementary to his (Mr Moore's) paper on the same subject,* and showed the progress made in the improved modes of constructing wrought-iron pipes since the date of that paper. In his son's absence he offered the following remarks on the discussion.

Mr John Thomson clearly stated the point upon which the whole question as to the adoption of malleable iron or cast-iron pipes turned—viz., the question of cost.

On the question of cost of manufacture the price of materials and of wages were so different in San Francisco from those in Glasgow that the cost there would form no criterion of what it would be here, but the cost of making the pipe might be taken at nearly the cheapest cost that sheet or boiler plates could be manufactured into any shape with a proper plant, properly systematised. It could be made with the cheapest labour and to a great extent with special tools, so that the labour on wrought-iron pipe and on cast-iron pipe could not differ very much. The working load was nearly 4 to 1, so that if the cost of wrought-iron pipe per ton was four times as great as cast-iron, it would still be as cheap per foot. Of course a larger margin for

---

* See Vol. XXVI. Transactions of the Institution.

safety is required for cast-iron, and the engineer must determine which was the most suitable for the work, but he thought they would be found to be cheaper than cast-iron.

He was of opinion that the lasting properties were largely increased by the coating, and thought too much stress could not be laid on this point. At the same time the duration of uncoated pipe would astonish any one who had not seen them. He knew of pipes that had been more or less in constant use very close on 20 years. He had sent for samples of the asphaltum and would forward it to the Secretary. With regard to lead joints without the inside ring, as shown in the diagram, he had made many joints without the ring, but where a good joint was required the ring was better and really cheaper. It kept the ends of the pipe perfectly fair. It did not contract the opening as the ring was not smaller than the inside course, and it kept the lead from running into the pipe while the joint was being made; besides, if the pipe was not to be placed underground, the ends might be kept a little apart to allow for expansion in case of the sun striking it while empty. The Risdon Iron Company had made many such pipes for use in the Australian gold mines, shipping them in prepared pieces from San Francisco.

With regard to leakage. It was very important that no spouting leak should exist where the pipes were buried. He had seen a pin hole leak be the means of cutting through No. 9 iron in a few hours by the eddy of the water and sand. This showed how tight the 25 miles of pipes referred to in the paper were, as no such leak occurred.

With regard to corrosion in different soils. A soil called "adobe," a black loam, had a bad effect, and when soil of this kind occurred the pipes were laid in clay or sand.

With regard to Mr Howard Bowser's remarks as to the fear of caulking rusting off and the pipes then leaking, his experience of pipes of this kind was that the coating filled every space, and the caulking kept the asphaltum from being pressed out. He also thought if the leak was small the floating impurities of the water would fill up the spaces.

With regard to riveting the round seams, there were many miles of pipe with no riveting in the round seams and no caulking. These pipes depended on the dirty water keeping them tight, the floating impurities getting into the spaces. He thought if the coating was properly put on there could be no rust. The coating was a bed of asphalt—good but thin.

With regard to Mr C. C. Lindsay's remarks as to the inside ring diminishing the water way, he had already pointed out that the ring was not smaller than the inside course and there was, therefore, no contraction. Various kinds of joints had been tried, but the joint illustrated had given great satisfaction with pressures up to 1000 lbs. on the square inch. On pipes from 12 inches to 42 inches diameter it could be made with little trouble. In making bends the inside ring steadied the ends and the pipe could be moved to any position and made ready for the lead joint.

With regard to Mr James Watson's remarks as to the lasting properties of these pipes, he had already made some remarks on that point. He had stated 20 years as the length of time they had been in use. Many pipes had been in use a longer time, but he had given the length of time that pipes of a large diameter had been in. These pipes were still running, and were in good order. He saw no reason why they would not be in use for other 20 years. The water was from a rain-catch and was nearly the same as the Glasgow water. As regards repairing. When a leak did occur it was usually a spot left badly protected with coating, and was seldom more than a small spout. This was stopped by putting on a piece of soft rubber and strapping it up against the leak with a thin band and screws. It could be done in a few minutes. In case of a joint pulling apart by the settling of the ground a lead joint could be run round it in the same manner as if two pipes were jointed, or a length could be cut out and the ends jointed in the usual way. All partial leaks could be permanently patched with india-rubber. All this work at joints was done by the ordinary skilled labour—not boiler makers, but men accustomed to laying water pipes.

He was much pleased that the paper had given rise to such a

discussion.   It was a subject worth investigation.   He could assure them these pipes had been of immense service in San Francisco, and it was a question for engineers to decide whether they were not applicable here.   If there were any further point where his experience would be of service he would be glad to explain it as far as he was able.

On Erecting the Superstructure of the Tay Bridge.

By Mr ANDREW S. BIGGART, C.E.

(SEE PLATES IV., V., AND VI.)

Received and Read 23rd November, 1886.

Two sessions ago I had the honour of laying before this Institution, a paper dealing shortly with the mode of sinking and building the substructure of the new Tay Viaduct, and noticing the many original devices, adopted by Mr Arrol, for the carrying out of that work.

To-night, I will confine myself to the Superstructure, and at once turn to it, merely premising that it will be found that the methods adopted vary as much from the ordinary modes of erection as did those which were employed in the preliminary parts of the undertaking.

As the mode of erection employed at the south end of the Viaduct, necessarily varied considerably from that at the north end, and still more that at the centre of the river, from either of the former, I shall deal with the erection of each of these three parts separately, thus endeavouring to bring all the salient points under your notice.

In the paper already referred to, the description left off at that part of the piers where the iron base was securely fixed to the brickwork, by means of strong bolts, built well down into the same.

This base is composed principally of channels, the upper ones being of an octagonal shape, and suitably formed to receive the wrought-iron superstructure, which is securely riveted to it. About eight feet above high water the two bases over each pier are joined to one another by other channels, which form part of the connecting piece, they being built into it. All the bases are similar in design, although they vary much in size.

THE SOUTH END OF VIADUCT.

Here the wrought-iron piers (see Fig. 1, Plate IV.) have an exceedingly graceful form. From each of the channel bases, two octagonal columns spring, which become united as they near the top. This connection is easily effected, by curving the three inner sides of each octagonal column till the two columns meet and form an arch immediately over, though high above the connecting piece of the piers; the two adjoining sides of each column being meanwhile widened on the flat till they also meet each other in a common plane. The three remaining, or outer sides of the columns, are carried to the full height, and thus leave the pier, at the top, an irregular though eight-sided figure. The sides and ends of the pier batter towards one another as they near the top.

The whole of the pier is plated, the thickness of the plates varying from $\frac{7}{16}''$ to $\frac{1}{2}''$. At each of the corners, two splayed channels form the outside connection, while inside they are connected by an obtuse angle. Both outside and inside tees cover and stiffen the vertical joints of the plates, while horizontal diaphragms are introduced, at regular intervals, to stiffen the whole pier. Near the top of each pier four short girders are securely riveted, to carry the bedplates of the main or roadway girders.

On the premises at Glasgow, where all the work preparatory to erection was executed, the whole of the piers were put together before being despatched, and at this stage a large part of the riveting was done. They were then taken down and forwarded to Dundee for final erection. To facilitate this re-erection, care was taken that the joints of all the angles, plates, channels, &c., were, while giving ample strength, so arranged as to allow the piers being taken down in large sections.

To provide a steady working platform, from off which the work of erection might proceed, one of the pontoons, which had previously been employed in sinking the cylinders of the bridge was floated alongside the pier, and in the manner formerly described, carried above the influence of the tide, by its supporting columns. On this platform a derrick crane was placed, of a height sufficient, and so fixed, as to command the entire pier until completion. The various parts of the pier were, as required, brought alongside the pontoon in boats. Under these conditions the erection was carried on so expeditiously, that an entire pier could easily be built in six days. After erection, the unfinished riveting was completed. This was done by hand, the men working from off a rising platform, composed of two light lattice girders, running along each face of the pier, the platform being meanwhile suspended from four small winding drums, fixed to two portable frames resting on the top of the pier.

Communication is established between the old bridge, and these piers (see Fig. 1, Plate IV.), by means of small rope gangways, consisting of two curved or carrying ropes, a few vertical standards, with crossbars for carrying a timber footway, and two upper ropes, used principally as hand rails. The ends of the ropes are securely fixed to the old and new bridges, and are tightened by means of simple union screws.

As soon as the first pier was completed, the girders required for the southernmost span (the whole four of which are new) were built on a projecting platform, secured to the old bridge, and from it slid to their final position in the new bridge, on beams, stretching from the old to the new structure.

After several piers were completed, arrangements were made to have the roadway girders of the old bridge removed intact, so as to form part of the new. This was most successfully accomplished, and with perfect ease and safety.

Two large pontoons (see both views marked Fig. 2, Plate IV.), each 80 feet by 27 feet 6 inches by 8 feet 6 inches, designed for this purpose and the subsequent floating out of the large 245 feet span centre girders were employed, with certain special appliances. These pontoons were securely bound together, at a parallel distance of 20 ft. from each other, by heavy cross girders. The arrangement was as follows:—Firmly fixed to the outer edge of each pontoon, and securely braced to one another, were two wrought-iron columns, rising to a great height. In each of these columns was placed another or telescopic column, so fitted as to be easily raised or lowered by a hydraulic ram when required. This hydraulic ram was secured to the fixed column, and had a stroke of about 7 ft., but by arranging different points of attachment to the sliding column, a lift of 13 ft. could be obtained. To secure temporarily the sliding column to the fixed, and thus make the two as one for the time being, a series of equally pitched pin holes were provided in both columns, in line and opposite to one another, through which pins might be passed when desired. On the top of each pair of columns a strong girder extended from the one to the other. These girders formed the support for the roadway girders, while the spans of the old bridge were being transferred to the new structure. If these two pontoons, and their various parts, be looked upon as one complete pontoon, then the size would be 80 ft. by 75 ft. ; dimensions sufficiently large to at once impress your mind with certainty as to the absolute safety, so far as stability was concerned. Each section of this pontoon was supplied with steam boiler, winch, capstan, bollards, and all other appliances necessary for mooring and handling such a craft. Hydraulic pumps were also on board to supply the power required to work the rams, used in raising and lowering the columns and their load. The full load the pontoon was designed to carry safely was 600 tons.

Thus equipped, this pontoon could in less than half an hour lift from the old bridge and transfer and place on the new, a complete span, consisting of two roadway girders, and the bracing between them.

The plan adopted in the removal was as follows :—Near the time of low water, the pontoon was placed immediately beneath the girders to be removed. If the difference between the height at which they were, and that to be occupied by them, was great, then the hydraulic rams, within the columns, were put out well nigh their full stroke ; with this, and the variable height to which the sliding columns could be telescoped, it was so arranged that a space of a few feet was left between the roadway girders and those on the top of the lifting columns.' Hardwood packing was now placed on these cross girders, immediately underneath the four points where the two roadway girders would take their bearings. But a short time now elapsed till the rise of the tide enabled the pontoon to lift the whole span from off its resting-place, and thus leave both free to be slowly carried along by the current, as the mooring ropes were slackened away till the load was adjusted over its new seat. This being accomplished, the whole four sliding columns were simultaneously lowered. Thus the girders of the little used roadway of the former Tay bridge are made to form part of the new, although this first position is not their final one. It is determined by the use to which they are now put, viz., to form a roadway on which to run out the new, or additional centre girders, and from off which they are to be lowered into position. These new centre girders, required to form a double line of rails on the new bridge, were built, and riveted almost complete, on the high ground taken up by the railway sidings, immediately to the south of the bridge. Full advantage was taken of the numerous lines of rails, for carrying the material to the place of building, and also for travelling the cranes along, during the building and riveting operations.

The girders were built parallel to the lines of rails, care being taken to have them in the same relative position to each other as they would ultimately occupy on the bridge. The riveting was done by hydraulic machines. These were hung from a light crane, temporarily placed on a railway truck, running on any convenient line of rails, commanding the entire girder. The whole of the girders were riveted up singly, and were in all cases at a considerable distance to the side of the line of rails leading directly to the bridge. This consequently entailed, when built, their being moved sideways till in direct line. For this purpose a special traveller was designed (see Fig. 3, Plate V.), being so constructed as to be capable of being moved forward as well as sideways, so that after having moved the girders to a convenient position it might run them out on to the bridge. After a trial of this method, however, it was found more convenient to move the girders sideways by a traveller designed for this purpose only (see Fig. 4, Plate V.), and then run them on to the bridge by the one just mentioned, and from it lower them into position.

The traveller used for the side movement was of the simplest form. It consisted of two box girders placed at right angles to and immediately over, the bottom boom of the girder to be moved, each box girder being provided with two running wheels. Against a bracket on the outer ends of the box girders there butted movable wooden struts lying at-an inclination to one another, and carrying overhead a short cross beam capable of supporting the top boom of the girder. These struts were braced to each other, and in this fashion combined the whole into a strong though temporary carriage. Two of these carriages were used for transporting a girder, each of which was placed about one-third of the span from its respective end. As soon as the box girders were in place, two crossheads were, at each carriage, passed under the bottom boom of the girder to be moved, and through these and the box girders bolts were inserted, which, on being tightened up, raised the girder from the ground and transferred it meanwhile to the carriages. These were now travelled over temporary rails till in position with the line leading directly to the bridge. Here the girder was lowered and the carriages employed thus far removed to make way for the other carriages which were to run on to the bridge. These were the carriages referred to (see Fig. 4, Plate V.), and were of a form somewhat similar to those already described, though made wholly of iron, and in every way of a much more substantial build. Each carriage was provided with eight wheels, four to each box girder. The inner wheels were set to the ordinary 4 feet 8½ inch gauge, while the outer ones were pitched to the gauge of the girders of the old bridge, by this time transferred to the new structure. On these old girders temporary rails had been laid to form the necessary roadway. Fixed to the crosshead of each of these carriages was a hydraulic cylinder provided with links for lowering the girder into its final position on the piers. These hydraulic cylinders and links were those formerly used for lowering the foundation cylinders, and were fully described in the paper I have already referred to. As soon as these carriages were in place, an engine was attached, which slowly moved all forward over the then completed portion of the new bridge, until it arrived at the point where only the girders of the old bridge formed the roadway. At this point the second set of wheels came into play to the relief of the first. If more than one span had to be gone over, then the carriages were drawn forward into position by the engine moving backwards, while attached to tackle extending beyond the furthermost point to which the girder had to be drawn. After arrival at this point special pipes were attached to the hydraulic cylinder, and then the girder, in a similar manner to that employed in the case of the foundation cylinders, was lowered on timber blocks previously placed on the top of the piers. In the case of the first lowered it was moved a little across the piers to allow the second new girder to be placed alongside in a similar way to the first. After this had been accomplished the whole four girders (two old and two new) were separated until they rested immediately over their final position on the top of the piers. They were meantime carried by temporary end posts and plates (put in position before being brought out), to allow of the flanges and webs, which formed them into continuous girders, being riveted up in the places where necessary. While this was proceeding the floor plates and the bracings connecting the girders to one another were put in place and riveted, so that when these several parts were completed the bed plates on the piers could be at once placed in position and the four continuous spans finally lowered thereon. This last lowering was accomplished by means of a hydraulic ram acting on the end of a long lever which was inserted between the two bedplates on the piers.

The work on the south side is rapidly approaching completion, being a repetition, in its various stages, of the work just described.

NORTH END OF VIADUCT.

On the north, or Dundee side, of the Tay little difficulty was experienced in placing the work in position, this being specially the case where that work was on land. The whole of the piers and girders, out to the skew arches, were new, and were erected by means of a steam crane travelling along the old bridge. From the second span beyond that point the outside girders in the new bridge were those transferred from the old.

In several cases the iron piers are simply plated boxes resting on the channel bases already mentioned. These boxes are flat-sided, the sides and ends having a varying batter similar to that of the piers already described. The remainder on to the centre piers may be said to be those of the south end in miniature. Like them they were also made in sections of a convenient size for handling during erection.

The erection at this point was accomplished by means of a crane, fitted on board a small steamer, which also served to carry out the material to the pier. A single pier could be completed in the working part of four tides, which represents about 16 hours' work. After erection, any riveting requiring to be done was completed, and then the old girders, spanning the opposite space in the old bridge, were removed to their position in the new structure. In effecting this removal in a considerable number of the spans, near the north end, a different method from those formerly mentioned was employed. At the one end of the girder to be transferred (that end being the one next the centre of the river) a strong trussed timber was securely fastened to the pier and stretched across to the corresponding pier in the new structure. As soon as the bracings, between the girders, were cut away, the end of one of them was gradually slid, by means of a crane stationed on the old bridge, along the beam, the other end meanwhile being moved, by means of a crane in a boat, and slowly moved by it, towards its new position. After being conveyed thus, the end supported by the boat crane was transferred to one on the new bridge, the other end remaining as before. The gradual sliding was continued till the whole safely rested on the new bridge. When a girder had been thus removed, the companion girder was similarly shifted to its new position, although in neither case the final one, because they had now to serve as the roadway on which the new, or additional girders required, were to be run out and lowered into place. So soon as the pontoon could be conveniently employed, the girders were removed by it, and the remaining work completed in a manner similar, and with the same appliances, as were used at the other end of the bridge.

CENTRE.

A different mode of erection than those just described had to be adopted for the central or large spans of the bridge. These spans are opposite to that part of the old structure which fell on that eventful night in December, now almost seven years ago. Unlike the smaller spans the girders here are not continuous. In overtaking this section of the work, each span was finished on land, floated out to the piers, and laid in position on them, while at a low level, and afterwards raised to the full height by hydraulic appliances.

The girders, floor and bracing, comprising the span, were built on a specially constructed timber jetty (see Fig. 5, Plate IV.), placed at a point convenient for handling the material, and afterwards floating the whole span off by means of pontoons. The jetty was provided with two docks, of a size sufficient to admit the pontoons, and into which they could be floated at right angles to the girders. During the time the girders were being built, these docks were temporarily covered over so as to form a continuous platform on which the building and riveting might proceed.

The girders were built and riveted by hydraulic power, in a manner somewhat similar to those already described. In this case, however, the trough shaped floor, resting on the bottom booms of the girders, and the bracing connecting the two girders, were riveted while in position, thus completing the span. Two spans were simultaneously built, and completed alongside each other. So soon as any two spans were finished, preparations were made to float them out (one span at a time) to their respective piers. These preparations consisted in placing on each pier, upon which the girders were to rest, timber blocking, to carry them to the proper height, and also the fixing of chains and other tackle to assist in

9

bringing the girders into true position. At the jetty, the timber covering over the docks was removed to allow the two pontoons to be floated underneath the first of the spans. This was done at the ebb tide, immediately previous to the flow, during which it was taken out. The pontoons (see Fig. 6, Plate VI.) were those which had been employed in transferring the girders of the old bridge to the new, but were necessarily stripped of the upright columns. In their stead, however, timber cradles were placed at the four points where the span took its bearings. After having been floated into the docks, and the necessary tackle attached, nothing further had to be done but await the rise of the tide. During this interval, the whole system had to be carefully watched, in order to make certain that the girders were taking their proper seat. After this had occurred, but a short time elapsed for to effect the transference of the full weight of the span from off the jetty on to the two pontoons. When sufficient clearance was had, the pontoons, with their complete span, were then gradually pulled clear of the jetty by means of the steam winch on board each pontoon. This was generally about an hour before high water. Hawsers from four steam tugs, lying close by, were now attached, and a start made to tow all well out into the river. As the jetty was on the east side of and close to the old bridge, the tendency of the tide was to carry the pontoons towards it, and in towing the aim kept in view was to go well down as well as out into the river, and while there, turn the span broadside towards the bridge, opposite to the space which it was to occupy when in place there. The tide would then gradually carry it to its position, being in its course guided by the tugs. Several times two of the tugs had to be transferred to the upper side of the span, to overcome the effect of a head wind, too strong for the current caused by the tide. In all cases the spans were soon brought between the piers of the old bridge, and at once tackle extending from the old piers, as well as the new, was attached to the pontoons, to assist in guiding the spans to their position, over the timber blocking, on which they rested on the piers. On this being attained, screw chains, from the ends of each girder to the opposite sides of the pier, were now brought into play to accurately adjust the span at each end. At the same time, and indeed until the tide was at its full height, additional timber was added to that already on the piers, to keep under the space between the girders and the blocking. When the tide began to recede the girders very soon took their bearings; this being hastened by the opening of the valves, and admitting about eighteen inches of water in each pontoon. So soon as the weight was wholly transferred to the piers, the 'pontoons floated away, and were immediately towed to the jetty, to be in readiness for similar duty during the following tide.

On the material and plant being brought forward, a start was made to build the piers around and above the ends of the girders just placed. This was done by a crane placed on a trestle resting on the floor of the girders, and able to command the whole pier. It was built in large sections, in a manner similar to that employed on the south side. Here, however, owing to the ends of the girders passing through the sides of the pier, large portions were, for the present, left unbuilt. The part of the pier immediately over the top of the girders, however, was built complete; and this, with the additional. strutting put in under the arch, served to stiffen the incomplete pier, and ensure the safety of the succeeding work, and the safe lifting of the whole span to its final position. Several sets of lifting columns, hydraulic jacks, and cross girders were employed (see Fig, 8, Plate V.). While lifting any but the end spans of these centre girders, four columns were used at each pier. This gave one column to each end of the girders. In order to increase the rigidity of the two columns on either side of each pier, they were braced to each other and the pier. To make the description clearer, I will confine myself to a single column, and the work performed at it. Each column rested on a sole plate directly over the channel base of the pier. It consisted of eight steel angles (8 in. × 4½ in. × ⅞ in.) placed in pairs, in the form of a rectangle. The deep member of these angles was at right angles to the main girder, while the 4½ inch member was placed to the outside, to receive more freely the bracings running parallel with the girder. Other bracings above and below the girder, but at right angles to it, were secured to the deep members by bolts passing through the large pinholes in these members, drilled for the primary purpose of securing the cross girders to the columns during the after raising operations. During the lifting there were required underneath each end of the main girder three cross girders and one hydraulic jack. The cross girders were formed of strong steel plates with stiffening angles. The plates extended to and passed through between the deep members of the angles of the columns, and were secured by means of steel pins passing through them and the angles. The hydraulic jack was fastened to the under side of the main girder, on the centre line of one side of the column. Underneath and to it was secured one of the cross girders. The other two cross girders were placed in the other half of the lifting column, the upper one being fixed to the main girder. Upon the latter would rest one quarter of the weight of the span, on the timber being removed from under the end of the main girder, thus allowing the load to be transferred to the cross girder.

The lifting of the span was carried out by raising each end alternately; the power being supplied by a set of hydraulic pumps placed on the floor of the span they were raising. At first only a little water was admitted, until the ram took up one quarter of the weight of the span, and relieved the upper of the two cross girders The pins inserted through this girder were then withdrawn. Water being again admitted the girder began to rise, taking with it the hydraulic cylinder and the upper cross girder. Between the upper and lower cross girders, hardwood packing was constantly inserted as the girder rose, in order to prevent any lengthened drop should anything give way. The span having been raised $7\frac{1}{2}$ inches, the pin holes in the upper cross girders, and in the column, were again in line. At once the pins were inserted, and the main girder allowed to take its bearing again on the cross girder. The pins passing through the other two cross girders were now withdrawn, and these girders raised the same height as the main girder. The girder next the hydraulic jack was raised by it, while the other was lifted by means of a simple screw.

As a rule, two of these lifts of $7\frac{1}{2}$ inches were made at one end of the span before it was raised at the other. The lifting is so continued till the span has reached its full height. During the lifting, much time is taken up in removing the plates, &c., of the pier, immediately over the girders, to make a passage, and in placing in position and riveting those underneath. To fix the lifting columns to the piers, and afterwards replace the bracings of the columns, is also a work of considerable time. The raising accomplished, the ends of the main girders are carried over the top of the piers, by a girder passing between the columns, till they are removed, upon which the remaining permanent cross girders and bedplates are placed in position, and the main girders lowered by means of a hydraulic jack thereon. On removal the columns are transferred to the next span about to be lifted; there to repeat similar work to that already performed. Two spans are usually raised at the same time, although each may be at a different stage. When raised to the full height, and lowered on to the pier, these large spans are practically ready to receive the permanent way, for, as we have seen, the whole was riveted up complete before being floated out.

At the three different sections of the bridge, the work is thus being rapidly pushed to completion, and now we are within a measurable distance of the time when communication by rail, over the estuary of the Tay, will be re-established; this time over a structure the graceful and scientific design of which will be a lasting monument to Messrs Barlow & Son, the engineers of the undertaking, and in the execution of which the genius of Mr William Arrol has shown with such lustre. As regards the quality of material and workmanship, nothing is left to be desired. Add to this the fact that a wind pressure of 56 lbs. per square foot has been provided against, and that each foundation has been satisfactorily tested to one-third more than the greatest possible load which could be put upon it, and we may, with every confidence, assert there is not the remotest possibility of the fate of the first Tay bridge being that of the second.

The discussion of this paper took place on the 21st December, 1886.

Mr T. A. ARROL said they must all thank Mr Biggart for the valuable papers on the Tay Bridge which he had brought before them. He had looked over the paper carefully, and he thought it reflected more than credit upon Mr William Arrol for the manner in which those works had been carried on. The skill and foresight displayed by Mr Arrol was something extraordinary, and in those works they saw an example well worthy of being followed. There was no scruple about expenditure : it was no mere matter of pounds, shillings, and pence ; but there was an evident desire to have the greatest security for the men employed, and to get the work done in the best way, and so as to give satisfaction to all. For him to criticise the paper he felt would be folly—in fact, presumption ; and he would only further say that he was very much indebted to Mr Biggart for the paper, and to Mr Arrol for permitting so much information to be put on record with regard to this great work that he had undertaken.

Professor JAMES THOMSON said he was in cordial agreement with the speaker who had just sat down, in his thanks to Mr Biggart, and in his admiration of Mr Arrol, for the great ingenuity and foresight displayed, and so successfully carried out, in the execution of the Tay Bridge.

Mr JOHN THOMSON said he had great pleasure in reading the paper. He had recently the advantage of going over the Tay Bridge with Mr William Arrol, who showed him all the details, from the beginning to the end. He was of opinion that it was a paper of great value, and that this Institution was under a debt of gratitude to Mr Arrol for his liberality in so freely and openly showing all the minute details regarding the bridge's construction, for the benefit of the members of this Institution in particular, and of engineers in general. He thought the thanks of the Institution were due to Mr Arrol and Mr Biggart for the pains they had taken in bringing the matter before them.

The CHAIRMAN said he quite agreed with what Mr John Thomson and other gentlemen had said, in thanking Mr Biggart for bringing forward this very interesting paper. It was one thing to design a bridge of that magnitude, but it was a very different thing to construct it and carry out the work. He believed that it was easier to design it than to construct it according to the design; and a great many of the difficulties arose, not from the design, but in the erection of it over such an expanse of water, and on the estuary of a river so much exposed to wind, and where the currents were so rapid, he believed sometimes equal to six miles an hour. Mr Arrol had overcome those difficulties very well.

On the motion of the Chairman, a vote of thanks was passed to Mr Biggart for his paper.

On Improvements in Valves for Steam Engines.

By Mr JOHN SPENCE

(SEE PLATE VII.)

Received and Read 21st December, 1886.

THIS description has reference to a new or improved construction of oscillating, or it might be rotating, valve, and its casing or chest, and arrangement of the parts and passages of ports of the same, suitable for shutting off the flow or pressure of fluids, or the reversing of these to and from motive power engines, or for other purposes where these operations are required to be performed.

By one arrangement, suitable for the reversing valve of ordinary steam engines, the oscillating moving part of the valve, A (see Figs. on Plate VII.) is formed with a disc working face plate, A′, with two or more opposite segments, $a\,a'$, cut out of its outer periphery corresponding to similar ports, $b\,b'$, formed on a circular face plate, B′, of the valve chest, B. The one port, b, in the chest, B, leads to the inlet division, c, of the pipe, C, and the other to the outlet division, c', of the pipe, C, in the valve chest, B, leading to or from the engine. The one quadrant port, a, is cut quite out or through the disc, A′, of the valve, so that when left over one or more of the said ports, b or b', in the face plate of the chest, admits the steam or pressure fluid from the valve casing, D, above and around the valve, down into the inlet port, b or b', of the stationary valve face and division of the chest, B B′, and pipes, C, while the other or opposite quadrant port, a', in the valve face opens into a close

10

exhaust chamber on the back of the valve, leading into a central circular port, A^2, in the valve face, and in the stationary working face, B', of the valve chest, B, down through an exhaust central pipe and branch at B^2 in the chest, led by any outlet pipe desired, as to the atmosphere, or to a condenser when applied to condensing steam engines. The valve chest, B, with its central or circular exhaust port and outlet pipe, B^2, would be formed with a deep annular outer chamber, b^2, all round the central pipe, B^2, below the working face, B', into which the inlet and outlet ports, b b', would enter this annular outer chamber, b^2, having two longitudinal ribs or divisions to divide it into two separate chambers, b^2, b^2, each formed with a separate lateral branch and divisions, c c', of its own, the one leading to the inlet port or ports, and the other from the outlet ports of the valves of the engine or engines, as the case may be. The turning of this quadrant or segmental ported disc valve, A, a half turn (or quarter turn, as the case may be), on or over the said reverse segmental ports, b b', in the working face, B', of its chest, B, reverses the direction of the steam (or other acting pressure fluid), changing what was previously the exhaust outlet ducts and ports of the chest into the inlet, and those which were the inlet now into the exhaust and outlet passages. Or by turning the valve a quarter round (or an eighth, as the case may be) in either direction from the said two open positions, this would shut both ports and passages in the reversing chest. The upper part of the chest, B, is formed with a broad circular flange, to which the corresponding flange of the close valve casing, D, over and enclosing the valve, would be jointed, having a lateral branch, E, for admitting the steam or other acting fluid all round the valve into the chest, B. On the back of the casing, D, a stuffing-box would be formed concentric with the axis and actuating boss of the oscillating or rotating valve, A, for passing its actuating spindle, F, through, and keeping it steam or fluid pressure tight. The spindle, F, of this valve could be actuated outside its casing and stuffing-box by a hand wheel, or it might be by a lever and rods from a distance, or by wheels when required to rotate. This valve is also capable of being made an equilibrium valve by making the area of its bearing rim, on back of valve, equal to the area of face of valve.

This stop or reversing valve, A, and valve chest, B, D, would be particularly applicable for leading the steam to and from the double-ported pipes, C, leading to and from the trunnions of oscillating engines; or for reversing the motion of rotary or reciprocating steam or other fluid pressure engines in which the reversing is effected by the changing of the inlet passages, *c*, into the exhaust, *c'*, and the exhaust into the inlet, and *vice versa* for reversing back again. This valve is also suitable for double cylinder horizontal engines, such as are used for winding and hauling purposes, and for cranes and winches, or for steering engines on board ship, all of which require to be frequently and quickly reversed, and all without other reversing gear than the use of this valve (see Fig. 7, Plate VII.). In such cases the slide valves for these engines would require to take their steam in the backward direction from the inside of the slide valves.

The discussion of this paper took place on the 25th January, 1887.

Mr S. G. G. COPESTAKE said he had not had the advantage of hearing the paper read. From the diagrams he supposed the valves described were round in form and worked in a circular manner. Now, he had a great objection to all such kinds of valves; and from looking at the drawings he was satisfied that these valves would thoroughly fail so far as steam was concerned, and also any other fluid or gas.

Mr DUNCAN (Whitefield) had looked into the paper, but did not feel that it called for any special remark. He had been visited by Mr Spence and learned that that gentleman also had used, or proposed to use, the compound system for the attainment of expansion. This was of more importance than the style of reversing-valve, as when expansion was got the range of application of valve-reversal was greatly increased, but there were some difficulties in the way.

The CHAIRMAN (Mr Dundas) regretted the absence of Mr Spence, who might have assisted in the discussion; and he would now close the discussion, and in doing so would propose that they award Mr Spence a hearty vote of thanks for his paper.

On an Improved Engine Revolution Counter.

By Mr MATTHEW TAYLOR BROWN, B.Sc.

———

(SEE PLATE VII.)

———

Received and Read 21st December, 1886.

———

THE author described Kaiser's Patent Revolution Counter, two forms of which were exhibited.

The general features of this counter, which he claimed as improvements on those counters generally in use, were :—

(1) The method of locking the number wheels after each successive movement (see Figs. 1, 2, and 3). This is accomplished by means of a projecting rim on the number wheel passing between the overhung teeth of the pinion, which latter cannot move till a gap which is in the rim comes round opposite the pinion, and so relieves it, a single tooth on the boss of the number wheel at same time gears into and moves the pinion one tooth forward, equal to one-tenth of a revolution.

(2) The higher number wheels, which have in ordinary circumstances to move very seldom, cannot get set fast by corrosion on the axle. The axles, both of the number wheels and pinions, in the Kaiser counter, are carried right through from end to end, and move with each stroke of the engine, thus keeping them free.

(3) The stroke of the driving lever does not need to be limited, as the motion is so arranged that the driving shaft may either simply oscillate, rotate forwards, or rotate backwards, the resultant motion of the number wheels being the same in each case.

The driving shaft actuates a small crank, the pin of which works in a slotted lever, and this slotted lever is secured to a segmental ring, to which it gives a reciprocating motion. The segmental ring in turn gives motion to a toothed wheel on the axle that carries the train of number wheels. The teeth of this wheel are shaped as in a clock escapement, and the motion is really a clock escapement reversed.

The discussion of this paper took place on the 25th January, 1887.

Mr A. C. KIRK said he really did not see that there could be much more said on the subject of this counter. This counter proceeded on the principle of making a one-toothed wheel drive one of several teeth. Ratchets and levers had been applied in counters hitherto, and generally they did pretty well; but unfortunately sometimes they slipped. He could not see that the teeth of that counter could miss, and so far as he knew the principle on which it acted was quite new. He thought they were all much obliged to Mr Brown for bringing the counter before them. He had no doubt that the principle of this counter might be applied to other cases, which would occur to engineers.

The CHAIRMAN (Mr Dundas) asked if Mr Brown could inform the meeting if the patent counter had been tried in Glasgow or neighbourhood?

Mr M. T. BROWN was sorry he could not answer the question fully, as though there had been some sold in Scotland they were not sold directly to the users, so he was unable to say where they were working. In England, however, a considerable number were in use. There was one on H.M.S. "Scout," also they have been fitted to the new sewage machinery being erected at the Houses of Parliament, where they have to register the number of sewage tanks discharged into the main, and they have been supplied to Messrs Armstrong, Mitchell, & Co., of Elswick, to Messrs Avery

& Co., of Birmingham, for use on automatic weighing machinery, and to many others.

The CHAIRMAN said this did not seem to be a subject productive of much discussion. Of course it would have been an advantage if they could have been referred to some engine in the neighbourhood; and to some practical trial for the sake of comparison with others generally used. He was sorry that there was so little to say upon this subject. He moved a hearty vote of thanks to the author for his paper.

On the Shafting of Screw Steamers.

. By Mr HECTOR MacCOLL.

———

(SEE PLATE VIII.)

———

Received and Read January 25th, 1887.

———

No part of the machinery of screw steamers gives more trouble than shafting badly designed, badly made, or badly looked after. But while this is so, the quantity of literature on the subject is in inverse ratio to its importance, and the following paper is therefore intended to record and discuss well known facts, not to propound new theories, nor to prove superior knowledge.

In screw steamers the shafting is divided into three main sections —viz., crank shafts, intermediate shafts, and propeller shafts, and we cannot do better than follow this order, with such subdivisions as may be necessary.

CRANK SHAFTS.

Material,—Iron.—Up to the present time by far the greater number of shafts have been made from forged iron, and the qualities which have secured for it this honourable position are its moderate cost, facility of manufacture, and character for reliability. The prices of solid forged iron shafts from the beginning of 1870 till the beginning of the current year are shown in the accompanying graphic table, Fig. 1, Plate VIII. This table shows the fluctuations which have taken place from time to time, and nothing can better exhibit the present deplorable depression of marine engineering than the unprecedentedly low price of crank shafts.

It would have been interesting to have shown in the same table

11

the prices of steel shafts, but owing to the great variety of steels used and corresponding difference in rates, it is impossible to institute any direct comparison between the price of steel and iron shafts, except for a very short period.

Facility of manufacture is almost implied in moderate cost, and is in this case due to the demand. At present an iron crank shaft of moderate size can be delivered in the United Kingdom in about 14 days, whereas a similar shaft in steel could not be delivered much under double that time.

But the quality which, more than any other, has kept forged iron in a front position, is its well deserved character for reliability; for although it is practically impossible to forge a flawless iron shaft the material is of such a nature that defects may be known to exist for a considerable time before they develope so seriously as to become dangerous, and rarely indeed does an iron shaft break without having given ample warning.

It is impossible to secure anything like homogeneousness in a large mass of forged iron, and the difficulty increases with the mass of material in the forging, and is greatest in the ordinary form of shaft where the webs, pin, and body are forged together and afterwards machined out of the solid, and it is doubtful whether, under ordinary circumstances, this form of shaft should be used for any but small sizes. A step in the right direction was made by coupling double throw crank shafts at mid-length, the forging is more manageable, the finished shaft easier handled both in the shop and in the ship, a fracture does not involve condemnation of both shafts, a defective pin or journal can be shifted into a position of less responsibility, and a half shaft is sufficient to carry as spare. A still further improvement has been made by returning to early practice, and the shafts, webs, and pin are forged separately, machined, and built up, as has always been done with the shafts of paddle engines; and by this plan only, can shafts, even of moderate diameter, be made of iron with any certainty of success.

The absence of homogeneousness in forged iron also shows itself in the difficulty of obtaining a smooth and hard skin on the bearing surfaces or journals of the shaft, the practical effect of which will be noticed further on.

Steel, hammered and unhammered, has been extensively used for crank shafts in recent years. It is undoubtedly superior to iron in strength, stiffness, and homogeneousness, and therefore it is the most suitable material for shafts 16 inches and upwards in diameter. Hammered shafts have been made from crucible cast steel, Bessemer steel, fluid compressed steel, and from Siemens-Martin steel ; and no doubt many of these have given satisfaction, but, on the other hand shafts made from each of these materials except the last, have failed, giving little or no warning ; and the proportion of such failures, as well as their repetition, has brought considerable discredit upon steel, and has given it a bad character for reliability. Many shafts have also been machined direct from steel castings, and no failure of these has yet come to the writer's knowledge.

The homogeneous nature of steel is admirably adapted to form smooth and hard surfaces on the journals, and therefore the average steel shaft runs with less friction than the average iron shaft.

But in spite of its many advantages, steel is only now beginning to make headway, the obstacles to its progress being to some extent its higher first cost as compared with iron, but principally its character for unreliability, of which instances will be given under another head. A good illustration of the practical result of this feeling of distrust is to be seen in the fact that in ordinary practice, notwithstanding its superior strength, no reduction in size is made in a steel shaft as compared with an iron one, nor would such reduction be considered prudent, in ships engaged on long voyages or in the North Atlantic trade. In special cases, however, where lightness is of great importance, or where full power is only used occasionally and for comparatively short periods, full advantage may be taken of the superior strength of steel ; and in one such case a shaft reduced about 15 per cent. has run successfully for $3\frac{1}{2}$ years.

Notwithstanding all that can be said against the material, there is little doubt that all objections to its use will ultimately be removed or largely reduced, and that steel will displace iron as completely in shafts as it has already displaced it in boiler construction. The determination of the best form of material to fulfil all requirements, rests with steel makers, but the writer expresses the opinion that at present the shaft which best combines efficiency with economy is a built shaft whose body and pin are forged from Siemens-Martin ingots of low carbon power, and webs made from forged iron.

Diameter.—The various strains upon a crank shaft and the dimensions necessary to resist them, are thoroughly investigated in Professor Rankine's treatises and elsewhere, and need not be discussed here.

Each engineer has studied the subject for himself, and by combining his knowledge and experience, has probably formulated in a way suited to his own practice, some method of fixing the diameter of a crank shaft. Nevertheless it is well that there should be some recognised standard of minimum diameter, and inasmuch as most new screw steamers either come directly under the supervision of the Board of Trade, or are specified to be constructed in accordance with its requirements, it has followed that the Rules published by that Department have been, in one way or another, largely used in this country.

The present rule, in a slightly modified form, is as follows :—

d = diameter of crank and propeller shaft in inches.

D^2 = square of diameter of L.P. cylinder in inches or sum of squares if more than one.

r = ratio of area of L.P. cylinder, or cylinders, to H.P. cylinder.

P = boiler pressure in lbs. $+ 15$.

S = stroke of piston in feet.

$$d = \sqrt[3]{\frac{D^2\, P.S.}{f(2+r)}}$$

$f = \begin{cases} 170 \text{ for 2 cranks, } 90° \text{ apart.} \\ 180 \text{ for 3 cranks, } 120° \text{ apart.} \end{cases}$

Diameter of intermediate shafting $= \cdot 95d$.

This rule is a vast improvement upon its predecessor, it stands upon a sound basis, and gives a fair strength of shaft.

Crank pin diameter and length.—In ordinary two-crank engines these are usually made equal to the diameter of shaft, as shown in Fig. 2, Plate VIII. Some engineers make the diameter greater and length less, with the view of obtaining bearing surface with reduced length of engine, and in some such cases where the crank pin gave constant trouble, its length was increased by thinning the webs on the inside, as shown dotted at AA in Fig. 2, Plate VIII., the alteration proving perfectly effective.

The proportions of crank pin shown in Figs. 2 and 4 give ample surface in three-crank triple compound engines, but in two-crank engines they should not be less than shown in Fig. 3, Plate VIII.

Many shafts are now made with a hole bored through the webs and pin, as shown in Figs. 2 and 3, Plate VIII., a steel or iron pin is supplied to fit this hole, and in the event of a serious flaw appearing the pin may be driven in and thus enable the vessel to complete her voyage in safety.

Webs in solid throw shafts are generally proportioned as shown in Fig. 2, Plate VIII., and are frequently made even thicker.

If the main bearings wear out of line, stiff, that is thick, webs are objectionable owing to their throwing all the bending action upon the neck of shaft or pin at BB, Fig. 2, Plate VIII.

It is therefore preferable to make them thinner than shown on Fig. 2, Plate VIII. ; and the proportions shown in Fig. 3, Plate VIII., give equal strength with more elasticity. These proportions also, with equal length of crank pin, bring the pairs of main bearings closer together, and nearer the line of stress.

Fillets at the junction of shaft and pin to webs are generally made 1 inch to 1½ inch, or even greater radius, with a view to strengthen these parts. But large fillets not merely shorten the bearings, and thus render them liable to heat, they spread them apart and remove them from their most effective position. To obviate this, the writer prefers a uniform radius of half an inch, believing that good bearings with small overhang are more important to the shaft than the doubtful strength of large fillets.

The fillet joining the shaft and couplings is sometimes made angular, as shown in Fig. 2, Plate VIII., but is more generally made as shown in Figs. 3 and 4, Plate VIII. But when an eccentric has to be fitted close to a coupling, the fillet should be unhesitatingly sacrificed to give the eccentric a good stable bearing on the shaft.

Webs in built shafts are usually proportioned as shown in Fig. 3 Plate VIII. In some cases, a long crank pin has been combined with closeness of the main bearings, by forming bosses on the cranks as dotted in Fig. 3, Plate VIII.; but unless the crank be very long it is impossible to overhaul the bearings in certain positions of the cranks, a fatal objection when, as frequently happens nowadays, a large steamer has to be "turned round" and despatched in a few days. In any case this plan renders it exceedingly dangerous to feel the necks of main bearings when the engines are in motion.

The allowance for shrinking should be about one-thousandth of the diameter in steel cranks, but possibly more may be required in those made of iron; this amount of shrinkage will hold well without overstraining either the eye or the shaft.

Some prefer not to key the crank on the shaft, but as it is quite possible for the heat from a warm pin to pass through the webs and slacken them on the shaft it is necessary that they be keyed, and a good plan of doing so is to bore one or two holes longitudinally along the junction of shaft and crank, and to drive in well-fitted steel pins.

The general adoption of three-crank engines has led to the designing of other forms of built shafts, many of which possess good features, which it is unnecessary to describe here.

Couplings are now always forged on the shaft, and their various proportions are shown in Figs. 2, 3, and 4, Plate VIII.

Some makers, with a view to getting the couplings true to each other during workshop operations, leave a projection on the end of one shaft fitting into a corresponding recess in the adjoining one, as shown at C, Fig. 2, Plate VIII. Such a device is quite unnecessary, but even if the contrary were true, the projections should be cut off before the shafts leave the shop, otherwise they cause great annoyance and delay when a shaft has to be replaced on board ship.

In actual cases fitted in this way the thrust block has had to be shifted aft enough to free the crank shaft, then both crank shafts lifted bodily out of their bearings before they could be uncoupled. In such cases the projections are speedily cut off, for when only 120 consecutive hours from the vessel's arrival in dock are allowed for the entire work of replacing a crank shaft and having the engines again under steam, all such novelties are dispensed with.

Bolts, for economy of manufacture and repair are moderate in number. They are fitted in three different ways, as shown at *a*, *b*, and *c*, in Figs. 2 and 3, Plate VIII., viz. :—

(*a*) Taper bodies with nuts but no heads.

(*b*) Taper bodies with nuts and heads.

(*c*) Parallel bodies with nuts and heads.

The first plan (*a*) has the advantage of being easily well-fitted in the workshop when made with a taper of about 1 in 24. It has, however, the serious defect that the couplings being held together only by the taper of the bolts, are liable to be drawn apart by the repeated reversal of the engines. In the case of a large steamer which grounded on a bar, her engines were kept running full speed astern for some hours, after which it was found that some of her couplings, fitted in this way, were parted from one-eighth to three-eighths of an inch.

The second plan (*b*) would probably fulfil all requirements, if shafting never had to be renewed ; but when a shaft has to be fitted on board ship the operation of rimering out taper bolt holes is neither speedy nor economical.

The third plan (*c*) is therefore that which, upon the whole, best meets all requirements. Good workmanship is required to give the exact amount of fit necessary to distribute the shearing strain amongst the bolts, and at the same time permit of their being easily backed out when required. If too good a fit, the bolt may either be destroyed in being backed out, or worse still, may have to be drilled out.

The plan, however, by permitting accurate templeting and rosebitting, greatly facilitates the rapid coupling of a new shaft, and this, after all, when combined with efficiency, is the important consideration.

The bolt head should be chamfered on the edge as shown on Fig. 3, Plate VIII., so that in driving up, the blow may fall in the line of bolt and not on the flange. The point should have a good projection to save the screw from damage in driving back. Both in driving up, and in backing out, a few sharp blows from a light ram are more effective than hammering, and less likely to damage the bolt.

Bearings.—Next in importance to the crank shaft are the bearings in which it runs. If these are insufficient in stiffness or in surface, are badly arranged, or are not carefully attended to, not only will those in charge have constant trouble, but the shaft itself will be flawed or broken, risking the loss of the ship, and causing her owners great expense.

Not only must good bearings be provided, but care must be taken in the design that the maximum working stress be equalised in each pair of bearings as compared with the others, otherwise unequal wear of the bearings may result—the most fruitful cause of broken shafts. The writer's experience is that with bronze bearings unequal wear is the rule in two-crank engines, and also that in such engines the forward bearings usually wear down faster, notwithstanding that the power may be fairly equalised in both cylinders. The reason of this is not far to seek, and the writer has long since satisfied himself that it is due to inequality of maximum stress. The following table, calculated from working indicator diagrams, shows the relative powers and stresses in cases taken without selection from different types of engine :—

Relative Powers and Maximum Stresses.

CLASS.		Cylinder.	Relative	
			Powers.	Maximum Stresses on Crank Pins.
Ordinary compound, 2 cranks.	1	H.P.	100	100
		L.P.	112·11	96·4
	2	H.P.	100	100
		L.P.	114·18	123·55
	3	H.P.	100	100
		L.P.	85·11	80·88
Ordinary compound, 3 cranks.	4	H.P.	100	100
		L.P.	82·95	136·52
		L.P.	91·66	151·3
	5	H.P.	100	100
		L.P.	62·93	61·65
		L.P.	63·74	61·65
Triple compound, 4 cylinders, 2 cranks.	6	H. & I.P.	100	100
		H. & L.P.	87·24	72·52
Quadruple compound, 4 cyldrs., 2 cranks.	7	H. & [2]I.P.	100	100
		[1]I. & L.P.	141·71	132·79
Triple compound, 3 cylinders, 3 cranks.	8	H.P.	100	100
		I.P.	109·41	107·87
		L.P.	102·25	106·69
	9	H.P.	100	100
		I.P.	133·6	154·49
		L.P.	135·39	125·57
	10	H.P.	100	100
		I.P.	104·66	131·06
		L.P.	100	130·31
	11	H.P.	100	100
		I.P.	101·52	102·83
		L.P.	93·53	84·39
	12	H.P.	100	100
		I.P.	100·42	174·78
		L.P.	104·65	124·56

It may surprise many to find that in this table the worst cases are to be found in three-crank engines of both types.

In designing new work excessive variations may be avoided, but as it would be unreasonable to expect equal wear in adjoining pairs of bearings working under stresses so unequal as some here shown, only the utmost care and attention in working, and in inspecting, will prevent fracture of shafts in such cases.

There is no doubt that in a well-designed triple compound three-crank engine, the crank shaft with its greater equality and greater subdivision of stresses, and with its more numerous points of support, will run better than that of any two-crank engine; and with good material, workmanship, and attention, such a shaft will last as long as any other part of the engine.

Material.—Good gun-metal, of a judicious hardness, if carefully fitted and attended to, will run exceedingly well, but if made sufficiently hard to resist wear will certainly fracture with a slight heat.

It is difficult to be uniformly successful with the mixture in gun-metal bearings; but if the truth were known, the skill and care of those in charge is less uniform in its quality than that of the material with which they have to deal, otherwise it would be hard to account for the widely-varying results shown by two or more sets of bearings cast at the same time and worked up from the foundry without selection.

Different kinds of bronze have been patented from time to time, and have been tried with varying success, but while each may have had good qualities, good gun-metal has, by its "all round" qualities, quite held its own.

Notwithstanding all this, the material called "white metal" is in one form or another rapidly and surely superseding the harder metals in bearings. Certainly no bronze could be trusted to perform the work which, voyage after voyage, has to be done in driving a modern mail steamer across the Atlantic with almost the regularity of clockwork. White metal has stood this, and many other tests, and it is capable of performing equally well in every grade of engine down to that of the small coaster. White metal bearings are generally looked upon with distrust by those taking charge of them for the first time, more especially when they have superseded bronze in carrying the same shaft; but after all this feeling leads to the exercise of greater care, with the invariable result that white metal, when fairly tried, becomes a friend for life.

A true shaft working in good white metal bearings certainly runs with surprisingly little friction, as proved by reduced lubrication and small wear. Both crank pin and main bearings have run several Atlantic voyages without losing the scraper marks on their surface. White metal bearings can also be run much closer than bronze, and this has no doubt something to do with reduced wear; and it necessarily follows that they occupy less time in closing and "leading," no small advantage in these days when vessels are so short time in port.

Crank pin bearings are fitted in two ways—viz., the connecting rod forged with a large end and bored out to receive circular brasses, or made with a flat butt and fitted with flat-bottomed brasses; in each case secured by a wrought-iron cap, and bolts through the butt of the rod. The former plan requires less brass, and the brasses can be finished in the lathe, but it neither presents such a good surface for taking the "knocks" that occasionally occur in the best regulated engine, nor is it adapted for lining up, while if the pin gets hot the bolts are bowed out so as to be difficult to withdraw, and if heating occur frequently the rod is almost certain to be found cracked across the crown.

Singular unanimity prevails amongst those in charge of machinery in preferring the flat butt rod, on account of the large surface rendering it easily adjustable, less liable to distortion, easily over-hauled, and readily felt while in motion; while notwithstanding the comparatively large amount of brass in it, the rod, as a whole, costs no more to manufacture, and can be renewed more quickly than the other.

White metal in crank pin bearings has generally been fitted in brass, and this is very convenient when the existing bronze bearing is in fair condition, for then it is easily altered to receive white metal. But where it has to be fitted complete, cast-steel is decidedly the most suitable material for holding white metal in crank pin bearings.

Main bearings are also fitted in two ways—viz., with the bottoms flat or semicircular.

If the bearing be bronze, it should be well-fitted in a flat-bottomed recess in the bedplate. The flange which generally projects round the edge of the brass should be cut away at the bottom to admit of liners being slipped underneath when required. The sides of the brass should be carried up to the centre line, so as to give as much side bearing as possible in the seat.

Round-bottomed brasses should not be fitted, as they cannot be lined up except at great expense, and are soon knocked out of shape. The only advantages they seem to possess are the small amount of brass in them, the facilities for machining them, and the fact that they can be removed and dressed up without lifting the shaft.

White metal, on the other hand, should always be fitted in round bottomed cases, for the wear is so slight that the bearing requires no lining up unless it has been warm, and in that case it should be refilled. If the bottom bearing be turned $\frac{1}{16}$ inch eccentric it can be readily removed without lifting the shaft.

In new work the most suitable material for carrying the white metal is cast-iron, for it may then be made as heavy and stiff as is desirable. Instead of having flanges at each end, one flange in the middle fitting in a bored recess is sufficient; the bearing is thus supported by the bedplate to its outer edge.

In bearings of moderate size the cap may be made of cast-iron or of cast-steel with white metal fitted directly in it.

White metal is fitted in several different ways, but the best is undoubtedly that on which it has been fitted from the first, viz., in longitudinal bars 3 to 4 inches wide, separated by strips about $\frac{1}{2}$ inch wide, the white metal standing about $\frac{1}{16}$ inch above these dividing strips when they are of brass, and about $\frac{3}{16}$ inch above them when they are of cast-iron or steel. The case for holding the white metal should, if possible, be tinned and heated to about the temperature of melting tin before the white metal is run in, and if these operations have been well conducted, little or no hammering will be required, but should hammering be necessary, it should be as regular as possible over the whole surface.

Some fit the metal in one large sheet, but the plan just described holds the metal better, and even if the divisions between the bars of white metal served no better purpose, they form excellent oil gutters, for retaining the lubricant against the journal, the benefit of which may be proved by cutting similar gutters in bronze bearings.

Lubrication, when effective, consists in the separation of the journal from the bearing by a film of lubricant ; and perfect lubrication would consist in supplying the precise amount necessary to effect this separation continuously. In the absence of perfection it is preferable to lubricate excessively rather than insufficiently, and in the first voyage of new machinery it is specially important that the journals and bearings be kept in good order, even at the expense of much oil, for this prepares the way for good working bearings with moderate lubrication.

The lubricant should be applied at as many points as possible, and it is a good plan to have the central oil tube in crank pin and main bearings so placed that it delivers in advance of the vertical centre line, thus ensuring the oil being carried on the crown of the bearing.

The alternations of stress in crank shafts are themselves favourable to lubrication, for without them it would be difficult to get lubricant on the side of a bearing directly opposite to a stress constant in direction. A very good illustration of this occurs in the case of a large and heavy crank shaft which tends to heat only when revolving very slowly, with no lift to admit oil to the bottom bearings.

Suitable oil gutters are essential to the proper distribution of the lubricant ; and in the case of white metal, these are supplied by the formation of the bearing as already described. In bronze bearings great fancy is often displayed in the formation of these gutters, but the main feature is to have them as large and direct as possible.

The oil tubes should be large in diameter so as not to choke readily, and those for the crank pin should have couplings so arranged as to permit them to be quickly removed and replaced.

In large engines, and in fast running engines, it is well to fit one or two telescope tubes to each crank pin, these not only ensure accurate hand feeding, but permit a slight heat in the journal to be readily detected.

The oil cups for crank pins should be fixed directly over the dippers; not on the cylinder lagging, connected to the dippers by long horizontal tubes, which retain the oil during the roll in one direction only to discharge it uselessly on the reverse roll.

The main bearings should be fitted with large oil tubes having cup mouths, over which should be fixed the oil boxes at such a height that each drop of oil may be readily seen falling into the cup mouth, and each syphon tube should have a small cock outside so that the oil may be shut off at once when the engines are stopped temporarily, and turned on again without delay.   The oil box should be mounted so that it may be pushed back to allow the withdrawal of the oil tube from the bearing, and care should be taken that the oil tube projects through the cap into the bearing, otherwise most of the oil may pass between the cap and the outside of the bearing.

The usual form of lubricator is the syphon, which by capillary attraction discharges more or less lubricant, according to the number of threads in the worsteds.   Many mechanical methods of lubrication have been designed, by which the oil discharged depends upon the revolutions made, and some of these work successfully.

The form of lubricator at present attracting attention is that in which the discharge can be regulated by a small valve and can be seen; such an arrangement promises well, but only extended experience can determine its true value.

Flaws and Fractures.—Iron shafts generally show their want of homogeneousness in "reed" marks on the surface of the journals, and few indeed are the iron shafts free from these.   Such marks generally show during the turning of the shaft, but they are then so skilfully "doctored" as to be almost invisible in the finished shaft, only to reappear after a little work. But the great resurrectionist of flaws is a hot bearing, for it not only re-opens those which have been carefully buried, but frequently brings to the surface others whose existence had been unknown.

When a "reed" is longitudinal, it does not affect the "life" of the shaft, but when it is of any great extent its ends are carefully centre-punch marked, and the "reed" is nicely flattened off with a smooth file. If it extends, a drawing or tracing of its form and position is made, on which is recorded its history and progress. If such a defect be serious, it sooner or later begins to run diagonally, and finally transversely across the shaft. A flaw which, when first seen, is diagonal or transverse is serious, for if of any extent it directly affects the strength of the shaft, not merely by reducing its effective section, but by presenting a weak point on which strains are more or less concentrated.

These flaws may be investigated, and their extent judged of in a variety of ways, and it is doubtful whether any other small detail of marine engineering inspection displays so much judgment and experience as the placing a correct value upon a flawed shaft. The simplest and easiest course is to condemn, especially when one has no direct interest in the shaft under judgment, and it frequently happens that where such shafts are cut up to ascertain the extent of their flaws, it is found when the information is no longer useful, that these were of an innocent character. It is, therefore, much more conducive to peace of mind to leave the study of condemned shafts entirely to the scrap merchant.

Fracture, through the crank webs, although common many years since, is now, owing to improved methods of laying on the slabs, very rare indeed.

Steel shafts are generally without mark or speck on their surface, and from the strength of the material it might be expected that if made sufficiently large in diameter they would last much longer than iron shafts. Such, so far, has not been the case, and in several instances which have come under the writer's notice they have not done anything like the work of the iron shafts which they had replaced, while in each case the flaw when first observed was of a serious character.

In one case a solid forged steel shaft replaced a similar iron one, and after a few voyages was itself replaced by a built steel one, which in turn had to be replaced by a solid iron shaft.

In other two cases, iron shafts which were considered rather light were replaced by expensive forged steel shafts, and in both cases the latter had to be removed within twelve months, and iron shafts again fitted. In each of these cases the flaw was transverse, at the junction of the shaft to the web.

In another case a crank pin 15½ inches diameter by 20 inches long fitted in a built shaft, after running four years, and having been warm several times, suddenly heated up, and water from the spray pipes was turned on; after running in this way for a few minutes, as the bearing showed no signs of cooling, the engines were stopped and the pin stripped. While the brasses were being dressed, and about half-an-hour after the stoppage, a sound was heard like a stroke on a small bell, loud enough to attract the attention of all around, and on examination a fracture was found extending in a straight line the whole length of the journal. The edges of the crack were filed down, the brasses put together, and the pin ran without further trouble for six years, when it was removed owing to the appearance of a transverse flaw. In this case the pin had probably been well heated through, and was of course cooled from the surface; then during stoppage the heat worked out again until it expanded the pin sufficiently to break the outer cylinder of cooled metal.

In all the cases of fractured steel shafts which have come under the writer's notice, the flaw has been so fine as to escape any but the most minute inspection, which is a dangerous feature in the material; and in each case the journal has been more or less heated, leaving the decided impression that the higher classes of steel are incapable of standing sudden cooling.

But irrespective of material, there is no doubt that the majority of crank shafts are fractured, not by defects from within, but by maltreatment from without, either through the journals not being true to each other, or through the bearings being out of line.

It is seldom indeed that the journals are so far out of truth as to directly endanger the safety of a crank shaft; the mischief is generally done indirectly by heating, and so unduly wearing, one or other of the bearings. The bearings may be out of line either through their having been so fitted, through insufficient stiffness in the bed plate or its supports, through weakness or flexibility of the structure of the ship, or through irregular wear of the bearings and journals. Whatever be the cause of this want of truth, if the bearings are untrue to each other, the result is a double bending of the shaft at each revolution, and if this be allowed to continue, fracture of the shaft at BB, Fig. 2, Plate VIII.

The limits of this paper do not admit of a detailed investigation into each of the causes of want of truth already mentioned; but as the last—irregular wear of bearings—is the most fruitful source of shaft fracture, it is necessary to enter into some detail regarding it.

The table on page 89 indicates one cause of irregular wear—viz., unequal stress in the different pairs of bearings, and consequently unequal wear; but so long as wear is legitimate, or is not accompanied by heating, the danger is not so great. But where bearings heat, and heating becomes chronic, with salt water lubrication, rapid and unequal wear follows as a matter of course, and then the fracture of a shaft is only a question of time.

Shafts which have broken through being out of line always show a fracture like that shown at BB, Fig. 2, Plate VIII., or what is known as a "ball and socket fracture." This is not a form of fracture which would be produced by insufficient torsional strength in the shaft, nor is it one which under any circumstances would occur at the neck of the crank if the bearings were always true.

The greatest care should, therefore, be taken, and the greatest watchfulness exercised, in ascertaining as frequently as possible the exact condition of the main bearings. Gauges are now generally

13

carried, showing the original height of the crank shaft at each bearing, and these should be regularly applied. Their testimony should not, however, be considered infallible, as the following case will show. It had been noticed on the outward voyage, that there was undue vertical motion in the forward main bearing of a large shaft, and on arrival in port, on being both gauged and "leaded," it was found apparently in good order. On the return voyage, to the surprise of the engineers, the motion was found to be as before. On the vessel's arrival at the home port, a rigid examination was made, when it was found that the shaft journal was ⅛ inch clear of the bottom bearing, and a liner of that thickness had to be inserted below the brass.

In another case, where the crank pin was very long, with an unusual stress upon it, and the main bearings also long and much overhung, the shaft appeared to spring a good deal at the necks, and yet was apparently true when gauged; but upon minute examination it was found that the long main bearings had worn wider at their inner ends, and thus allowed the crank to spring at work, while retaining its proper position when at rest.

No part of the engine-room routine requires more careful attention than the overhaul and examination of the crank shaft and its bearings; and, next to good management of the boilers, will attention to the shaft repay the time and trouble spent upon it.

We now come to the second division of the subject

INTERMEDIATE SHAFTS.

This includes all the shafting between the crank shaft and the propeller shaft.

Thrust Shaft is generally, and ought always to be, coupled to the crank shaft, and the thrust block should be placed in a recess in the after bulkhead of the engine-room, so that it can be readily seen and attended to.

Material.—Forged iron does well for shafts of small size; but in those of large size steel alone is admissible, being undoubtedly the best material for this shaft. The work being steady, and without shocks or blows, steel is well suited for it, while its smooth and regular surface in the thrust rings is greatly in its favour, as compared with the "reedy" and seamy nature of forged iron.

Diameter need not exceed that of the other intermediate shafts, or according to the Board of Trade rule, ·95 d. The couplings and bolts should be duplicates of those on the crank shafts.

Thrust Surface is now made much greater than was at one time the practice, being equal in square inches to ·5 max. I.H.P., or about 50 per cent. greater than what was at one time considered good practice. This large surface, which is that of the "horse shoe" collars, combined with the adjustable collars now universally adopted, gives excellent results, and the highest speeds can be run continuously with no more water than the drop necessary to form a lather.

The horse shoe thrust is, like many so-called modern improve-ments, an old invention, having been made many years since by Messrs Maudsley.

The number of collars varies with the power, and the following table gives some approximation of the number adopted in good practice.

Max. I.H.P.,	400	750	1150	1600	2100	2650	3300
No. of collars,	1	2	3	4	5	6	7
Max. I.H.P.,	4000	4800	5800	6900	8200	10,400	14,000
No. of collars,	8	9	10	11	12	13	14

The horse shoes are generally made of cast-iron for moderate powers, and of cast-steel for large powers, in each case faced with white metal.

The best arrangement for holding and adjusting them is a longi-tudinal bolt at each side of the block, made of manganese bronze or of delta metal, screwed from end to end ; each horse shoe being secured by two gun metal nuts. Where iron or steel bolts are used they are liable to rust, and so render adjustment difficult, if not impossible.

Bearings.—To ensure the thrust block being sufficiently long, it is well to cast a bearing on it at each end; each of these need not exceed d in length.

Lubrication.—The general principles already referred to should be carried out here, and, in addition, the block should form an oil trough, in which the shaft collars revolve.

Tunnel Shafts, in engines of moderate power, are seldom made of anything but forged iron, and although there have been failures, and bad failures, of iron shafting, these have been few indeed in proportion to the shafts running.

Diameter.—As given by the Board of Trade rule, is ·95 d, and this need never be exceeded; but the couplings and bolts should be duplicates of those on the crank shaft.

The length of intermediate shafts in ordinary sized engines is about 20 feet, and although they have sometimes been made much shorter than this, it is questionable whether the increased first cost is covered by the advantage of portability on the rare occasions when such shafts have to be renewed.

The shaft next the propeller shaft is usually made just so long that when uncoupled and removed, the propeller shaft may be drawn in far enough to allow the removal of the propeller.

The propeller shaft is, however, liable to so many unseen dangers that it ought to be drawn in for examination each time the propeller is removed, and this is best done by coupling to it an ordinary length of shafting, which is generally long enough to allow the propeller shaft to be drawn in at least far enough to permit a complete inspection. A recess should be provided at one side of the tunnel into which the uncoupled length of shaft can be pushed.

It is very important that each length of shaft should be rough turned all over, so as not only to let it run true, but to expose defects.

Bearings are best made of cast-iron lined with white metal, there being one bearing to each length of shaft. These are generally provided with lightly-bolted shell covers, but one well-known engineer has removed these covers from the shafting of several of his steamers, leaving the journal open so that its condition can be seen at any time. There is much sound reason in this departure, for the strength of cap and bolts is obviously not designed to resist upward pressure.

The depth of the bearing block should be rather greater than half of the diameter of shaft coupling.

Each block should be planed on the bottom, and well-bedded on hard wood not much more than 1½ inch thick.

Lubrication is usually provided by means of the ordinary syphon arrangement when oil is used, with a box alongside in which any of the lubricating tallows may be used if preferred. Save-alls should be provided at each end, as much for cleanliness as for economy.

Flaws and fractures are almost entirely due to defects in the material, and fracture is most frequently in a diagonal direction, through a scarph.

Intermediate shafting is rarely found truly in line even after working only a short time, and want of truth may always be looked for whenever and wherever the nuts on coupling bolts show a tendency to slacken. It is rarely indeed that this want of truth is so great as to seriously affect the strength of the shafting, nevertheless many proposals have been made for applying flexible couplings to the intermediate shafting, but these are quite unnecessary, and inventors would be surprised at the extent a line of shafting may be out and yet run comfortably and well. Some ships are so flexible at the after bulkhead in engine-room that if the coupling bolts between the crank shaft and next length be removed the coupling will shift more than once during loading or discharging cargo, and here a flexible coupling might in such cases be fitted with advantage, seldom anywhere else.

While it is necessary that the line of shafting be truly laid out and fixed when the machinery is first put on board, it should not be forgotton that this provides for truth only under one set of conditions, and that the one under which hardly any work is done. No one with experience of shipping believes that the steamer has yet been built which does not wriggle and twist even in moderate Atlantic weathers, and yet although there is nothing but its own flexibility to meet these changes of line, the shafting usually out-lives two sets of engines. This would indicate that it is possible to go too far in taking up and re-lining shafting which has given no trouble, for the mere satisfaction of securing its accuracy in one isolated set of conditions.

No system of lining by means of a cord can give the true centre line of shafting, nothing but correct sighting can do this; but while the greatest possible care should be taken to have the line true at first, the test afterwards should simply be the ease or trouble with which the shafting works.

We now come to the third and last division of the subject :—

PROPELLER SHAFTS.

The propeller shaft is in some respects even more important than the crank shaft, for while the latter can be regularly inspected at small cost, the former cannot, as a rule, be examined oftener than once a year, and then only at considerable cost; and while much may be done in driving a vessel with a broken crank shaft, the breakage of her propeller shaft renders her machinery useless for propelling purposes. Therefore no pains should be spared in making the propeller shaft and its fittings as perfect as possible, and every opportunity should be taken to draw it in for examination.

Material.—The majority of propeller shafts have necessarily been made of forged iron, and as a material it has given satisfaction, failures being ordinarily due either to external circumstances or to defective workmanship.

No doubt the greater strength of steel makes it more suitable for shafts of the largest size, and although the failures which formed the subject of Mr Davison's paper, read before this Institution last year, might seem to some minds to cast suspicion upon steel as a material for propeller shafts, there is little or no doubt that by the adoption of suitable precautions such failures can be entirely avoided.

Here, as in crank shafts, the writer prefers a steel shaft forged from a Siemens-Martin ingot, as being upon the whole the material from which any but a very moderate sized shaft should be made, and the principal reason for this preference is that such a shaft is without weld or scarph.

Diameter of the propeller shaft is usually made equal to that of the crank shaft, and although in some cases it is made larger, this is not with much reason, for while it is less open to inspection and more liable to deterioration, it has on the other hand less work to do than the crank shaft, and more steadiness in the performance.

The couplings should be in every way duplicates of those on the crank shaft.

Bearings.—Since the expiry of Penn's patent, the great majority of shafts have run on wood strips with plenty of water lubrication.

The outer end of the shaft from a little within the fore end of propeller boss to the inner end of the outer bearing is covered with hard gun metal from $\frac{1}{2}$ inch to $1\frac{1}{4}$ inch thick, according to the size of shaft. The inner part of shaft from about one diameter beyond the stuffing-box to the inner end of gland is covered with a similar liner. Each of these liners is shrunk on the shaft, and is further secured by a few screws tapped through the brass into the shaft, and riveted over flush; this is to prevent an accident which has sometimes occurred when the outer liner, probably heated through working in sandy water, has worked back into the tube, leaving the shaft to thump about and knock the bearing, and possibly the stern tube, to pieces.

The outer bearing is formed of a brass cylinder turned a good fit into the stern tube, bored internally, and secured by a stout flange and tap bolts to the outer end of stern tube. It is filled with lignumvitæ staves $2\frac{1}{2}$ to 3 inches wide, which are prevented from moving circumferentially by two brass staves riveted to the tube, one at each side of the horizontal centre line. The junction of the staves at their inner circumference is bevelled off or grooved out to form water ways. Usually only one brass stave is fitted, but the other is a much superior plan, inasmuch as it permits of the bottom staves, where all the wear is, being removed and renewed without disturbing the upper half. End wood wears very much better than ordinary strips, but it is almost impossible to renew it without withdrawing the bush from the stern tube.   A brass ring pinned to the flange of bush prevents the strips from working outwards.

The inner bearing is usually made of hard gun metal, and extends from the bottom of stuffing-box about one diameter outwards.

The inner end of the shaft, between the stern tube and coupling, is usually carried on an ordinary bearing.   It is of importance that the diameter of shaft be increased in this bearing until it is not less than that of the shaft liners; this permits the shaft to be drawn in without removing the block, and the latter affords a convenient support to the shaft while being drawn in.   Provision has also to be made for removing this block without disturbing the shaft, and this is done either by parting the block vertically and bolting it together, or by seating it on a stool which can be removed and the block turned round on the shaft until it is free.   In the absence of some such precaution, the block when worn has had to be cut to pieces, and a new one in halves fitted.

The shaft is often made with a collar, which bears on the fore part of this bearing, and prevents the shaft from moving outwards when uncoupled.

Outer End of Shaft.—The propeller boss should be made with a recess about 1 inch deep at the inner end, and large enough to pass easily over the brass liner.   The other proportions of boss should be somewhat as shown in Fig. 5, Plate VIII.

Taper.—A great variety of tapers is used, but upon the whole a taper of about 1 to 12 is best, holding well, and not being too hard to start off.   Most shafts are made with less taper, 1 in 16 being common, but this is too fine, being liable to split the boss, and hard to start off; it also lets the boss a long way on the shaft when refitting has to be done.

Fit.—Whatever be the taper, the boss must at all cost be made a good fit on the shaft.   If the fit be not good, especially at the fore end, water gets in, and then both shaft and boss waste and work until the fit gets unbearably bad and the vessel has to be docked. It is then better to remove both shaft and propeller to the shop and fit them thoroughly, for it is almost impossible in ordinary circumstances to make a first-class fit of a propeller in the graving dock. The whole boss should be perfectly watertight on the shaft, and to secure this, as well as to prevent contact between the brass, iron, and salt water at the after end of shaft liner, a soft rubber ring is slipped on the shaft at the latter place and the propeller boss is screwed hard home on it, making that part thoroughly watertight.

Key.—There are many opinions about keys, but upon the whole, it is best to have one key running the entire length of boss; this is more readily made a good fit than two short keys, while providing equal shearing section. It should be well fitted at the sides, but free from the keyway on top, and its work will be easy, provided the boss be well fitted on the shaft.

Nut.—The diameter of screw is such as to be well below the small end of taper, and its length sufficient to give ample clearance, and a little "drift" for re-fitting. It is generally screwed the reverse hand of the propeller, so that if the key be sheared the propeller will tend to tighten up the nut; the thread should be V, and the nut made a good fit on it.

Stopper.—A great variety of stoppers are made. A stopper should not only be effective, but adapted for readily following up any little move that may be got on the nut. The one shown at DD, Fig 5, Plate VIII., appears to fulfil these conditions fairly well.

Flaws and Fractures.—The majority of propeller shafts fail either through defective workmanship, that is through imperfect welding, or through corrosion at the ends of the brass liners, or at the joints of these when made in more than one piece.

In the former case fracture frequently takes place at a scarph, when the stern tube is almost certain to be destroyed.

Corrosion is the principal destroyer of propeller shafts, and it appears to be caused by galvanic action between the brass, salt water, and iron or steel, the latter being the sufferer.

When the brass coverings on the shaft are made square at the ends, as shown at E, Fig. 5, Plate VIII., corrosion of the shaft, as sketched at E, is almost certain to follow, and if allowed to go on, the destruction of the shaft is only a matter of time. This corrosion forms a groove

14

round the shaft, more or less irregular in depth, and very frequently eccentric to the shaft's circumference. While broad and easily distinguished on the surface, the corrosive action dies out radially in a fine line, so that it is extremely difficult, in turning down a shaft so grooved, to get the corroded line removed without ruining the shaft; and in some cases, in order to reduce the shaft's section as little as possible, it has been necessary to turn the groove eccentric. Whenever a shaft is large enough to permit of the corroded part being cut out, the groove should be broad, with easy sweeps, as dotted at E, Fig. 5, Plate VIII.

The shaft between the liners is generally more or less corroded, but being distributed over the whole surface, this corrosion is seldom of serious extent. In the case of a large steel shaft, after four years' service, this part was hard, with a skin like enamel, with only a few scattered patches of corrosion of very small extent.

A variety of plans have been tried with a view to preventing, or reducing the amount of, this corrosion.

Probably one of the earliest was to cover the space between the two main liners with brass. This, however, introduces an additional element of danger, in the joints required at the various lengths. These are seldom watertight, and when this is the case corrosion is even more rapid than when the ends of the liners are left open; curiously, the finer the joint, the more rapidly does corrosion go on, somewhat as shown at H, Fig. 5, Plate VIII. Even when the liners are checked as dotted, corrosion goes on underneath, unless the joint be perfectly watertight. In cases of this kind, corrosion is easily detected by the hollow sound when the brass is hammered at the joint, and no one could feel safe in replacing a shaft which so sounded until he had ascertained its actual condition underneath the liner.

Some engineers cast the liners on the shaft, and so cover them all over the part exposed to salt water; but the difficulty of casting brass solid on iron is well known, and if water gets through blown holes to the shaft, corrosion is certain to go on underneath.

Others cover the shaft all over in detached liners, "burning"

—that is fusing—them together at the ends so that the liner is practically all in one piece. It is found that the expansion at the ends of the liners makes them sound hollow, and whenever this is the case, as already mentioned, doubts are raised difficult to dispel by aught but a removal of the covering.

The most recent plan of covering all over is to make the liner, from propeller boss to stuffing gland, in one piece, shrinking it on the shaft, or forcing it on by hydraulic pressure. There can be little doubt of the efficiency of this plan, but its cost must be rather extravagant.

But, after all, most engineers prefer a known and seen danger to the chance of a hidden one, and so the great majority of shafts are left uncovered between the liners.

The simplest method of reducing the amount of corrosion is to bevel off the ends of the liners, as shown at F, Fig. 5, Plate VIII., keeping the shaft well painted at every opportunity. By this means, combined with keeping a good current through the stern tube, taken from the bottom, as shown in Fig. 5, Plate VIII., corrosion may in most cases be kept under. Indeed, the writer believes the experiment of forcing a stream of water outward through the stern tube would be found to prevent corrosion almost entirely. This could be done by a small pump worked from the shaft itself, and in the event of the vessel having to go astern on a sand bank, would probably prevent sand from entering the tube, and by heating them slacking the liners on the shaft.

Another plan is to "serve" the shaft with marline. This is an old and well-known method of protecting the space between the liners, and although not practised so much as formerly, is said to give good results. The writer, in discussing Mr Davison's paper, stated that in one case the "serving" was said to have unwound at sea, and had burst the tube; but, on making further inquiry, he finds that this is not correct. What happened was that the marline had got unwound on the shaft, and so hampered the drawing of it, that most of a week was spent in the operation, with the result that the practice was given up in that fleet.

A method patented by the writer, and applied to more than fifty shafts, is shown at G, Fig. 5, Plate VIII. The object of this plan is to cut off contact between the brass and salt water and the iron ; and this, in the most recent practice, is done by fitting an iron ring on the shaft at the part to be protected, slipping a soft rubber ring on the shaft between the end of brass liner and the iron ring, forcing the latter hard up, so compressing the rubber watertight, and while in this position pinning the iron ring on the shaft. Most of these shafts have had the iron ring shrunk on the space between it and the liner filled with white metal, and latterly with tin. In some cases the iron ring had been attacked by corrosion, but in every case which the writer has examined, the shaft itself was completely protected.

Apparently the simplest and most effective method of fitting the propeller shaft is one which has had a singularly quiet existence for years past. No brass liners are fitted on the shaft, but it is turned parallel from the coupling to the taper part. The bearings at each end are simply white metal bushes, the iron shaft running directly in these. Melted tallow is run into the stern tube, filling completely the space between it and the shaft, and the job is complete. Nothing to cause corrosion, nothing to get loose and cause trouble, if the white metal bearings last, nothing could be better ; and it is to be hoped that during the discussion some definite information may be given as to results.

In some cases the after liner appears to have been considered by the makers too long to be made in one piece, and so it has been made in two, either butted or lap-jointed, as shown at H, Fig. 5, Plate VIII. Rarely indeed is such a joint water-tight, and consequently many shafts have been destroyed by this simple oversight. In one case, where the ends were butted, and the joint so fine as to be hardly discernible, the shaft was reduced three-quarters of an inch in about fifteen months. In a similar case, where corrosion was not so extensive, a groove about half-an-inch wide was cut round the brass at the butt, and in this was laid, with sufficient spring to ensure its watertightness, a strip of soft square rubber, with a long scarph at each end where they joined. The whole was then wound round with several turns of copper wire, which was afterwards soldered together in several places. After twelve months' work, this was examined, and some of the strands of wire were found loose, but the job had been thoroughly watertight and effective.

The necessity for a certain amount of completeness introduces into such a paper as this much that is well known, and even elementary; and its principal value must therefore consist in inducing members to record their differing deductions drawn from observation of the same facts.

In the after discussion,

Mr E. KEMP said he thought it would be judicious to defer the discussion of the paper to another evening, and until the members had an opportunity of reading the paper. Mr MacColl had given them such a long paper and had travelled over such a wide field of time and material, that it would take considerable attention and pains properly to digest the paper. Mr MacColl had a rich fund of information, gained by experience and otherwise, so that he had no doubt that when they came to read over the paper, and to look into the many points that were treated of, it would be found to be of great interest and elicit a lively discussion afterwards. He proposed that the discussion be postponed till another evening.

The CHAIRMAN (Mr Dundas) quite agreed with Mr Kemp that the discussion on that very interesting paper of Mr MacColl should be adjourned until another evening, when they would all have had time to peruse the paper in a printed form, and when those who were able from experience to speak on the subject would have an opportunity of taking part in the discussion.

Mr JOHN KENNEDY said that the paper would require a great deal of consideration, as it was a long one and touched upon a large number of points of interest. As he would not likely have an opportunity of taking part in the discussion on another night, he would like to say a few words then. In regard to the question of flaws in shafts, his usual way of discovering a dangerous flaw was by simply passing his finger over it. If it could not be detected in that way it showed that it was not working As to longitudinal flaws they thought little of them. In protecting the propeller shaft, the practice he had adopted for a good many years was to serve it with marline. Then before bushing they painted it with pitch, which acted as a preservative and covered the marline. Instead of using lignum vitæ he had latterly used ebonite, and had found it stood for five years. He had had bearings of it running ten years, with only one-sixteenth of an inch of wear. With reference to crank shafts, he had used phosphor bronze very largely. Some bearings put in ten years ago had given no trouble. He did not use it now so much, as it had become very uncertain.

Mr D. C. HAMILTON had very great pleasure in listening to the paper, and he fully agreed with the proposal to postpone the discussion till another meeting; but he was of opinion that one night would not suffice for the discussion of the whole of the paper. With reference to Mr MacColl's remarks about propeller shafts without brass liners and running on a white metal bush, he had come across a steamer fitted in this way which had run seven years with the original white metal bush. They had occasion to cut out the liners and got others from the original builder of the boat, which they fitted in. To their surprise these did not last as many months as the former had done years. It was naturally thought that they were not made of the same material, but the builder asserted that they were the same, and explained that the fault lay in not putting in tallow. They got another liner and filled the tube with tallow, and some years after the boat was sold and the liner remained all right.

Mr S. G. G. COPESTAKE said locomotive engineers were so hampered for room that they could not compete with the great spaces of marine engineers in crank shafts. His sympathies were with malleable iron—good Yorkshire iron—for crank shafts. Although of late years steel had come to that uniformity of quality that one could get a good mild steel crank axle, still he was inclined rather to choose Yorkshire iron for that purpose.

The CHAIRMAN said the discussion would now be adjourned until another evening, when, he had no doubt, that this interesting paper would provoke a very lively discussion.

The discussion on this paper was resumed on the 22nd February, 1887.

Mr W. L. C. PATERSON said that those interested in the subject of this paper would feel gratified by the very able, comprehensive, and concise way in which the author had laid before the members the best means of designing and maintaining in efficiency the shafts of screw steamers. He had been struck by the similarity of views enunciated in the paper and the present practice on the Clyde. The writer had brought up many old facts for discussion; and though he agreed with nearly all that Mr MacColl had said, he might hold different opinions on some points, which they were all the better for thinking over and discussing. Speaking about crank shafts, Mr MacColl very appropriately told them, that the solid shaft was passing away, and that the built shaft would undoubtedly be the shaft of the future. It was now about seven years since the Anchor Line had adopted the built shaft, and at that time they gave the webs a shrinkage of about a thousandth part of the diameter of the shaft; but with the experience since gained they had increased this to about a five-hundreth part. On a crank pin which became loose they gave the new pin a three-hundredth part of shrinkage, which was, in his opinion, the right shrinkage, and that allowance had worked well to this day. He observed with a little surprise some remarks about the material of which the bearings of shafts were made. At page 90 of the paper Mr MacColl says :—"It is difficult to be uniformly successful with the mixture in gun-metal bearings; but if the truth were known, the skill and care of those in charge is less uniform in its quality than that of the material with which they have to deal, otherwise it would be hard to account for the widely-varying results shown by two or more sets of bearings cast at the same time and worked up from the foundry without selection." Now, he thought this statement showed some want of knowledge of the attention given to the shafts of ships in general. He spoke from experience and he thought that if engine-builders, and designers of the machinery would give more facilities for feeling the bearings it would be better. For instance, take a crank shaft where a bearing was so placed that it was impossible for the hand to get near it, as they would find the web of the crank at one end, and eccentrics or coupling bolts at the other end of the bearing, and so they had to put the oil in methodically, and wait the results. There was also a tendency to run the shafts at too high a speed. He thought that when they went over 60 revolutions per minute, they could not tell whether the shaft was being injured by heating or not, and hence had to trust entirely to methodical oiling. He observed that Mr MacColl was a great believer in white metal; and so was he (Mr Paterson); but he was of opinion that if brass were used in the same careful way as white metal, there would be fewer complaints against it—that is, by putting the brass in in strips, or cutting grooves through it. Or if they drilled a lot of flat holes into the solid brass bush, he was satisfied it would run very well after that, for the lubrication would be better. At page 98, the author says :—" Thrust Shaft is generally, and ought always to be, coupled to the crank shaft, and the thrust block should be placed in a recess in the after bulkhead of the engine-room, so that it can be readily seen and attended to." He did not see that they should accept Mr MacColl's view as being the best place to put the thrust block. Why should the thrust block be placed in a recess in the after bulkhead of the engine-room? He thought this thrust bearing so placed acted as a kind of fulcrum, for the crank shaft, as the structure on which the thrust block was fitted, was strong and rigid. It would be admitted that there was a good deal more of up and down wear on a crank shaft than there was on a thrust shaft; therefore, when this circumstance was considered it would be seen that they were apt to throw a greater strain on the thrust shaft bearing in this position than ought to be, hence the flaws so often found on the aft crank shaft bearing and pin, and on the thrust shaft. He would suggest that the best place for the thrust block was close to the propeller shaft, where they could make a strong and economical building, and thereby get rid of this strain. He was not satisfied with the way that propeller shafts were now fitted. The corrosion that went on was a source of great anxiety; and Mr MacColl very properly advised that every opportunity should be taken to draw in the shaft for examination. He had a notion that those shafts should be simply plain steel or iron, with no brass fittings, and made a quarter-inch larger at the bearings, and wood strips might be fitted in forward and aft in the cast-iron stern tube for the shaft to run upon. It might also help to prevent corrosion to fit on an oil syringe to force oil between the bearings once every four hours or thereabouts.

Mr MOLLISON said he was sorry that he had not had an opportunity of hearing the paper read, but from glancing over it, he did not see very much but what most engineers would agree with. He observed in the discussion that followed the reading of the paper that Mr D. C. Hamilton referred to a case in which brass liners had been done away with, and a white metal bush introduced. He had a propeller shaft drawn lately after running eleven years in white metal bushes, and without brass liners, and it was found to be quite clean, and bore no signs of corrosion. The stern tube had been filled up with tallow, which, in fact, had formed a bush the entire length of the tube inside. He did not see why this method should not be more generally adopted. With regard to the corrosion taking place at the ends of the brass liners on propeller shafts, that was a thing too well known to all of them, and various means had been tried to prevent it. Mr MacColl had introduced white metal rings at the ends of the brass liners, but he (Mr Mollison) had not heard how this had done. Others had lapped them with spun yarn or marline, which had kept them quite dry, and prevented corrosion.

Mr D. C. HAMILTON said he found it very hard to criticise the paper under discussion, and he must agree with others who had spoken, who said they were of the same way of thinking as Mr MacColl. It would be very hard to say anything against the paper, or even to ask questions upon it. One of the points in it was the amount of contraction given to built shafts. Mr MacColl divided a steel shaft into a thousand parts, and allowed one of these parts for contraction. Now he did not see anything different between a steel and an iron shaft, and he understood that Mr MacColl had iron webs fitted on steel shafts. Then with reference to the use of bronze in crank shafts Mr Kennedy, at last meeting, said he had tried phosphor bronze largely, and had been very successful; but had abandoned it, and gave the reasons why. He believed that where phosphor bronze was genuine it had always been very successful. His firm had steamers running for nine or ten years with it, and they were as good as the day they left the builder's hands. Mr MacColl said that if gun-metal was carefully fitted and attended to it would run exceedingly well. He believed that also. His firm's oldest steamers with gun-metal had been running 13 years satisfactorily, and were still doing well, never a hitch having occurred with them. He believed also that if they got good white metal it would also give good results; but he supposed that bad bronze, bad white metal, and bad brass, and gun-metal could not be expected to wear well. He was of opinion that if gun-metal had been kept up to the standard it had 25 years ago, white metal would not have been so universally adopted as it was. Mr MacColl in the paper said that in some of the Atlantic, and other hard-worked steamers, white metal had stood the tear and wear where no other material would stand. Mr MacColl spoke about the forward bearing wearing faster than the others. That was the case, and there was no disguising the matter. It was also said to be the case with the after bearing, but still the initial blow in the forward engine was always greater, coupled with the fact that the forward bearing was usually shorter than the others. He was very much astonished with the results given on the table at page 89—showing the difference between the initial powers and stresses. He thought it was within the capability of engineers to have the powers and stresses equalised. It did not say much for the scientific engineers to have such a difference between the powers and the stresses as the table showed, but Mr MacColl might be able to say that the engines were very well arranged both in regard to strains and powers.

Mr JAMES MURRAY said he thought this long paper might have been curtailed. As Mr MacColl acknowledged in the last paragraph of it, there was a good deal of the elementary. There was one feature in it, however. It was capable of drawing out the discussion of the members, more especially upon the corrosion of propeller shafts. He was glad to observe this, as it followed up Mr Davison's paper of last session, and which he thought had not received the attention it deserved. It was most important that any of the members who were able to suggest a remedy for corrosion in propeller shafts should come forward and give the Institution the benefit of their knowledge, as corrosion caused great depreciation to the property of shipowners. One or two points in this paper had struck him a little. At page 85 Mr MacColl suggested that fillets at the junction of shaft and pin to webs, might be reduced from one inch, and one and half inch, to a uniform radius of half an inch. He found in his own experience this had been a great evil, for engine builders had a certain proportionate length that they made crank pins, and with a view of strengthening the same they made a large fillet, forgetting that they were running away with the bearing of the connecting rod brass, and perhaps taking away a twelfth of the whole surface, thus impoverishing the bearing with the view of strengthening the shafts. In p. 86, Mr MacColl referred to the fact that some makers left a projection on the end of one shaft fitting into a corresponding recess in the adjoining one, so as to cause less trouble in fitting up. That was a very great evil, but he had not come across it except in one case, where they had one length of shafting to remove, and were not aware that such a projection was in existence, so that when they tried to get it out they could not, either at the one end or the other, and a diver had to be got to go down and cut off a *lignum-vitae* bush to get the shaft carried back.   He did not think there was many of these projections left on shafts now.   At page 91 Mr MacColl remarked :—" White metal bearings are generally looked upon with distrust by those taking charge of them for the first time, more especially when they have superseded bronze in carrying the same shaft ; but after all this feeling leads to the exercise of greater care, with the invariable result that white metal, when fairly tried, becomes a friend for life." Well, he could assure Mr MacColl that white metal would not be his "friend."   It was a "friend" that he would keep a jealous eye upon. So far as white metal was concerned in main shaft bearings it had done well ; but as regarded crank pins his experience led him to say that he would not use white metal for them.   He had no objection to what was said with regard to the high pressure crank pin brass, but it did not apply to that of the low pressure engine.   Instead of white metal he would advocate a brass bush, or a bronze bush, as some of the members had suggested.   Mr MacColl spoke of lubrication on page 94, and sought to attract notice to "the form of lubricator at present attracting attention is that in which the discharge can be regulated by a small valve, and can be seen."   This was nothing new, for he had seen it working 18 or 20 years ago. It was the proper thing for the lubrication of engines where it could be got at, and was very extensively used by many companies.   On page 97 Mr MacColl gives the breakage of crank shafts.   He thought Mr MacColl there came very near the point in the matter. by tracing it to insufficient stiffness in the bed-plate or its supports, through weakness or flexibility of the structure of the ship, or through irregular wear of the bearings and journals.   He believed that all sea-going engineers would bear him out in that opinion, and he (Mr Murray) had found it to be the case.   He had seen in one case that the flexibility had been so great that the shaft was broken two or three times, but after being stiffened he did not know that any other break took place afterwards.   He hoped that in this discussion the important matters treated off in this paper would be well considered, and he was satisfied that if all the members who are able would take part in it, they would all be thankful to Mr MacColl for bringing the paper before them.

Mr D. C. HAMILTON remarked that as to what Mr Murray had said about the high pressure and the low pressure pins, from his own experience he found that the low pressure pin caused most trouble. He thought that could be easily remedied by a slight modification of the slide valve. He believed that Mr Murray would have no difficulty of this kind in the future.

Mr GEORGE RUSSELL said there was one point, on page 101 of the paper, where the author stated that he held it was not necessary to apply flexible couplings to connect intermediate shafting, that he would like to notice. In December, 1874,* he had read a paper before this Institution, " On Marine Engine Shafts," in which he proposed to couple together the various lengths forming the screw shaft, which would allow the shaft to work perfectly free, although the line of bearings would deviate considerably. He still thought that was the direction in which they ought to go to lessen break-ages and friction in shafting. There must be a great waste of power, as well as a risk of danger, in the shaft being out of line, and he thought that some such couplings should be used between the different lengths of shafting. He believed with Mr Paterson that the strain would be somewhat less if the thrust block were further removed from the crank shaft than was the case in the present practice of engineers. It was a pity that more of the marine engineers were not present to give their opinions and experi-ence on the points raised. Both Mr MacColl and Mr Paterson had gone the length of taking out patents in connection with pro-peller shafts ; and it would be well if the Institution could get some information upon the practical working of their patented improve-ments.

Mr GEORGE OLDFIELD said he had had no experience with marine engines, but there was a point on page 83 of the paper with reference to the use of steel, that he would like some information

---

* See Transactions I.E.S., Vol. XVIII.

upon. It is there stated that steel is better adapted for bearings of shafts of 16 inches and upwards in diameter than wrought-iron, owing to the less amount of friction in running. His experience with Bessemer steel was quite the reverse of that. He found that bearings of wrought-iron in engineering tools gave better results, so far as friction was concerned. With reference to the table on page 89—relative powers and maximum stresses. Among them examples were given from the indicator diagrams of triple compound engines. He had always had the idea in connection with marine engines that there was a considerable amount of care exercised in designing this class of machinery, but it appeared to him, from the table, taking No. 12, that the result was not very satisfactory. For instance, he found a difference of 74 per cent. in the maximum stress on crank pins in the same engine, the power being about equal. They had first 100, and then the table jumped to 174·78; and then came down to 124·56. These discrepancies he thought must strike one with considerable surprise. Mr MacColl might say whether the examples that were given were from old, or from recent practice. It struck him that there should not be these great differences of stress upon the crank pin, and he could not understand how such should arise.

Mr GEORGE RUSSELL asked if it would not be practicable to have a crank shaft with only two bearings, one at each end. This would relieve the shaft from the strains generated by difference of wear, when three or more bearings are used.

Mr D. C. HAMILTON said that Mr Mackay could give them some information on engines that were being made with one bearing.

Mr EDWARD MACKAY replied that Mr Hamilton had remarked that marine engines were being made with only one crank shaft bearing. There were a number of engines of this kind at work, but without further experience in their working than they had had, he would not venture to recommend them. So far, experiments made in this direction, had not been altogether successful, particularly with engines of the larger size.

Mr WILLIAM CROCKATT would like to make a remark on one

point with reference to crank shafts. Mr MacColl had said it was advisable to keep the fillets at the junction of the shaft and pin to the webs *small*, so as to get as large bearing surfaces as possible, and to support the shaft close up to the neck of the web; but he would ask, Why should they not carry out the same principle in built shafts, and dispense with fillets altogether? He observed that the shaft shown on the diagram increased in diameter where it entered the webs. Now that seemed to him entirely useless. Certainly it was so in the pin, and it brought unnecessarily into built shafts the evils that were unavoidable in others.

Mr BRUCE HARMAN said with reference to white metal for stern bushes, it was found that it was the only material that would stand in the upper waters of the Humber. The water at Goole was very sandy, and though lignum-vitæ and other materials had been tried, white metal had been found to be best suited to meet the effect of the sand between the bush and the shaft; and had been used successfully for many years.

Mr H. MACCOLL, in reply, said they all appeared to agree with the various recommendations in the paper, and if that were so, he could say they did not practise what they knew to be right. Referring to the remarks of Mr Paterson and Mr Hamilton on built shafts, he had never made a built shaft for a screw engine, preferring to buy them finished, but the information as to shrinkage to be allowed in crank webs was given to him by the late Mr Leece (of Sir Joseph Whitworth & Co.'s), who had great experience in steel shafts, but owing to the softer nature of iron that material might stand a greater shrinkage, and no doubt the amounts given by Mr Paterson were correct. He was surprised to find that Mr Paterson proposed to fit the thrust block close to the propeller shaft, as he thought it would be generally admitted that it would be unlikely to receive proper attention if placed so much out of reach of the engineer in charge. In connection with the thrust block, he might say he recently examined one which, owing to its extreme shortness in proportion to its depth, had broken off the whole of the after flange at sea, showing the necessity for ample length, as pointed out at page 100. He did not think Mr Paterson's proposed method of fitting propeller shafts would succeed, indeed it was an old plan which had given considerable trouble. He was pleased to learn from Mr Mollison that the plan of running the propeller shaft in white metal, with the stern tube filled with tallow, had given such excellent results, and there appeared to be little doubt that that was the best of all methods of fitting this shaft. The suggestion to force oil through the stern tube had been on trial for about twelve months in a large steamer, a pump being used to discharge the waste oil from the engines continuously through the stern tube, and the result would be known shortly. He believed that the idea of an outward current through the tube was the correct one, and after the last meeting, Mr Hamilton had informed him that he fitted all his ships with a cock on the ship's side, connected to the stern tube, with a view to give this outward current. The curious course of currents about a ship's stern was shown by an ash ejector which discharged below the water having to be removed owing to the stern tube being found full of ashes. Replying to Mr Hamilton and to Mr Oldfield, the examples of powers and stresses on page 89 were all taken from new engines, and they were in no way selected. Nos. 8 and 11 showed that in triple compound three crank engines, both powers and stresses could be equalised; and no one who took the trouble to go into the matter carefully would have any difficulty in designing engines which would give equality in both directions. Mr Murray's opening remarks led them to expect some valuable information on the corrosion of propeller shafts, but he regretted that that expectation ended only in disappointment, as Mr Murray appeared to have nothing new to say on that subject. He did not think Mr Murray's experience of white metal agreed with that of those who had fairly tried it. Within the last few days he had seen crank pin bearings fitted with white metal, on which only one-eighth of an inch in the circumference of the nuts had been got, after several Atlantic voyages. Mr George Russell appeared to forget that in the discussion on his paper on "Marine Engine Shafts" in 1874, the necessity for flexible couplings was denied by most of the marine engineers present. The breakage of tunnel shafts was very rare, and the fact that these shafts were so lightly held in their bearings indicated that they were not broken through being out of line. With regard to his own patent coupling, he had never applied it to tunnel shafting, and had never asked any one to apply it. He had, however, applied it in several cases to the coupling of the shafts of centrifugal pumps to the crank shaft of the engine driving them, and here its utility was shown by the coupling running true at first and afterwards gradually running out of truth, showing the effect of unequal wear in the two shafts. In reply to Mr Oldfield's inquiry, he could only say that the journals of steel shafts almost always showed a better skin than iron shafts, which clearly indicated less friction in the former. To Mr Crockatt he might say, that in built shafts there would be no objection to their making the fillets as small as they chose, but it would be, in his opinion, bad practice to carry the shaft through the web at the same diameter as in the journal; it was increased in the web so as to stand the reduction of section caused by cutting keyways, and thus equalise the section throughout. He would like to say that since last meeting he had examined one of the first propeller shafts fitted with the patent arrangement shown at G, Fig. 5, Plate VIII. This had been running four years, and the iron rings were corroded somewhat as shown in the shaft at E, but the corrosion did not extend to the shaft, which had been preserved intact, and the ring could be easily renewed. The space was filled with white metal, which appeared to have a corrosive effect on the iron ring, but he had no doubt the rubber filling would prove effective.

The discussion of this subject was then further adjourned till next meeting.

In closing the discussion on this paper, on 6th of April, 1887, the CHAIRMAN (Mr Dundas) moved a hearty vote of thanks to Mr MacColl for the very interesting and elaborate paper that he had presented to the Institution.

16

The late Sir Joseph Whitworth, Bart.

———

22nd February, 1887.   Mr CHARLES P. HOGG, Vice-President,

in the Chair.

———

The CHAIRMAN said that, with the permission of the meeting, before beginning the business of the evening, he would like to draw attention to the great loss the engineering profession had sustained by the death of Sir Joseph Whitworth, one of their most distinguished honorary members. It would be difficult to over-estimate that gentleman's share in the advancement of engineering, due to the extreme accuracy and precision of the machine tools he made. His precision and absolute accuracy in any thing he under-took were amply rewarded, and about 20 years ago he gave no less a sum than £100,000 to found those scholarships which bear his name. He felt sure that his memory would be long cherished by the profession which his genius so much adorned.

On Collision Pads for the Prevention of Loss at Sea.

By Mr Richardson.

Mr Richardson described, and illustrated by a model, his Collision Pads for Prevention of Loss at Sea. The chief object of this invention is to provide means whereby holes or fractures caused by collision or otherwise, may be quickly closed to prevent the inflow of water and the sinking of the ship. The apparatus is easily handled, and can be applied in a few seconds. The pad is in a roll similar to a window blind, and, when required for use, is suspended over the ship's side and allowed to unroll itself over the fracture, which at once prevents the inflow of water. It is then secured by the shield.

The apparatus has been in practical use, and its efficiency is now thoroughly established, several steamship companies having adopted it.

The pads are supplied in water-tight cases, to be kept on deck or on the bridge, ready for immediate use.

The invention is also admirably adapted for ship lifting purposes, a number of vessels having already been raised by this means in a very short time after applying the pads. In one instance, the pad was applied to a large laden steamer, which had been sunk through collision; the vessel was pumped out, successfully raised, and kept afloat for a week with the pad on until her cargo (about 3,000 tons) was discharged, when she was docked for repairs.

When about to be used, the pad is handled by two sailors ready for unrolling. This is effected by the pulling of a slip line which fastens a strap, when the weight of the pad causes it at once to unroll, the slip line being held by one of the hands.

When the pad is completely unrolled, and the top guy ropes made fast to bulwark, the hole is covered, the flexibility of the pad allowing it to be pressed against the ship round the hole, and any serious inrush of water is thus stopped. While this has been in progress a messenger line has been passed over bow of ship and worked aft, and is now to be attached to the tail rope to carry it round ship to be secured. The shield is now applied by passing the guide ropes of the pad through each of the eyelet loops of the shield before it is unrolled, by which means it unrolls so that the steel cross bars come in between the transverse stays on back of pad.

Both pad and shield being now in position over hole, are completely secured by guy ropes to bulwarks, tail ropes from pad and shield round ship, and binding ropes from bulwark over shield, and round the ship to opposite side.

On the Education of Engineers.

By Mr Henry Dyer, C.E., M.A.,

Member of Council,

Life Governor of the Glasgow and West of Scotland Technical College.

Received, and held as read, 22nd February, 1887.

THE motion of which Mr Weir has given notice—namely, that steps be taken to obtain a direct representation of this Institution in the University Court—having come under the consideration of the Council, it was felt to be desirable that an opportunity should be given to the Members of expressing their opinion, not only on that special point, but also on the more general subject to which it leads up— that is, the "Education of Engineers"—for the only reason why the Institution should desire to be represented in the University Court must necessarily be to insure that the interests of engineering education should be properly attended to. No doubt if our President had been here he would have dealt with the subject, for he has always taken a very active interest in everything relating to the intellectual and moral welfare not only of engineers, but also of the community generally, and his opinions would have been of the utmost value. He, however, is detained by business in a foreign country, and at the request of the Council, I have undertaken to lay before you a few thoughts which may at least form the basis for discussion. The present time is peculiarly appropriate for the consideration of the subject of the "Education of Engineers," for not only is there a prospect of a Bill for reorganising the Universities, but at the present moment the Glasgow and West of Scotland Technical College is being formed, and the Council of this Institution

17

has sent a representative to its governing body. The opinions of the Members should be of great assistance not only to him, but also to the other Governors, coming as they do from those whose professions are likely to be most affected by the arrangements which are made. It must be understood that I wish my paper to be suggestive rather than exhaustive, as time would altogether fail me if I attempted to enter into all the details connected with any scheme of engineering education. I thoroughly agree with the object of Mr Weir's motion, and in the course of this paper I will try to show some reasons why it ought to be adopted. Perhaps I ought to explain, however, that Mr Weir's action is quite independent of mine, and doubtless he will also tell the Members what induced him to bring forward his proposal. I need scarcely add that I have been speaking and writing on the subject of the " Education of Engineers" for a good many years, so that no doubt many of you know my opinions, but in any case it may be convenient to collect them and state them briefly for the purpose of beginning the discussion. It is of course understood that any criticism I may offer of existing arrangements applies only to the system and not to the persons connected therewith, and therefore that there is not the slightest occasion for any personalities either in the paper or the discussion. In matters, however, relating to the general arrangements connected with the education of engineers, it is the duty of all interested in the matter to speak out bravely the truth as they know it, and resolutely and unflinchingly to do their duty, whatever the consequences may be.

In the first place, we must know distinctly what we mean by the term " Engineer " before we consider engineering education, for unless we do so it is impossible to understand what sort of preparatory training is required by those who propose to enter the engineering profession, in order that they may be prepared to undertake their work in an intelligent manner. If we consult the dictionaries we are told that an engineer is "a constructor of engines; a mechanist; one who manages a steam engine; a person skilled in the principles and practice of engineering, either civil or military." The vagueness of this definition well illustrates the state of public opinion on the subject, for the same name is applied to the man who designs and carries out important works, and to him who looks after a donkey pump. The popular impression seems to be that the name "engineer" is derived from "engine," but the etymology of the word shows that the reverse is the case. The Latin word *ingenium*, or the equivalent old French word *engin* means an innate or natural quality, while another old French word, *s'ingenier*, means to set one's mental powers in action to solve any problem whatever. From an etymological point of view, therefore, the name "engineer" is wide enough to include every profession, and "engineering" may be (and very often now is) applied to subjects not usually included in the work of engineers. It would occupy too much time to trace the development of that work, and to show how from small beginnings it is now one of the most important factors in modern civilisation. The name "engineer" was originally applied only to persons in the military profession, and the distinction between military and civil engineering appears to be of comparatively recent date.

Into the consideration of the former of these I do not propose to enter, as the teaching of that department is now for the most part taken up by special institutions. Our attention will be confined to Civil Engineering, although a great many of the preliminary subjects are common to both divisions. The term "Civil Engineering" thus evidently included all engineering other than military, and may be taken to mean the art of constructing large works for civil purposes in the design of which the physical sciences are applied. Within recent years, however, the domain of the engineer has widened to such an extent that the name Civil Engineering is generally restricted to the construction of such works as roads, railways, canals, harbours, and works of that class, and even these are now too numerous for one man to practise with success, and the result is that one department is fixed upon as a speciality, and the others are simply subsidiary. For purposes of teaching, minute subdivision is neither possible nor desirable, and it might be sufficient if what is now usually called Civil Engineering were divided into two parts, in the first of which would be included all works relating to roads, railways, and structures generally; and in the second those in which the principles of hydraulics were chiefly applied; and subject to this division the various branches of engineering are fairly represented by the classification which I adopted in a College on the other side of the globe which I had the honour of having the chief work of organising and managing, viz. :—

    1. Road and Railway Engineering }   usually called
    2. Hydraulic Engineering   } Civil Engineering.
    3. Mechanical Engineering.
    4. Naval Architecture.
    5. Electric Engineering.
    6. Architecture.
    7. Chemical Engineering.
    8. Metallurgy.
    9. Mining Engineering.

And as agriculture is now so much a matter of applied science—that is, of engineering—we may without unduly stretching the meaning of the term add another division, viz. :—

    10. Agriculture.

The mere enumeration of these divisions shows how utterly impossible it is for one man either to teach or to practise them all. A great part of the preparatory training is common to each, and while strongly insisting on the necessity of as general a training as possible in fundamental principles, and in the elements of the various branches, I would also point out the necessity in the present day for concentration to a large extent on one special subject, as the most successful engineer is the man who knows something of a great many subjects, but who knows more than any other person about one subject.

The Charter of the Institution of Civil Engineers defines *Engineering* as " the art of directing the great sources of power in nature for the use and convenience of man, as the means of production and of traffic in states both for external and internal trade," and then proceeds to enumerate some of its applications, such as in the construction of roads, bridges, aqueducts, canals, river navigation and docks, for internal intercourse and exchange, and in the construction of ports, harbours, moles, breakwaters, and lighthouses, and in the art of navigation by artificial power for the purposes of commerce, and in the construction and adaptation of machinery, and in the drainage of cities and towns. This enumeration includes the chief works which were carried on by engineers at the time the Charter was granted ; but the Institution has long since gone beyond it, by recognising the fact that the field of engineering has immensely extended since that time, and now, instead of being one profession, it is really a group of allied professions. For our present purpose, I will define an engineer as one who has been trained in the theory and practice of one or more departments of constructive or manufacturing industry, and who is capable of designing and superintending large works of construction, or of directing large manufacturing establishments. This definition may seem to exclude a great many to whom the name engineer is now applied—namely, the foremen and higher class of workmen—but while by no means wishing to undervalue their importance, for the purposes of education it is necessary to consider them by themselves, which I will do further on. I may remark that in all the discussions which have taken place recently on technical education, in my opinion too much has been said about the education of workmen, and not sufficient about that of the masters or directors, as I think that the latter is even of more importance than the former. In the future, the discoveries which will create new industries, or revolutionise old ones, are likely to be made by men who combine the highest theoretical and practical training. The work done by Bessemer and Siemens in metallurgy, by Sir W. Thomson in telegraphy, by Young in the manufacture of oil, would have been impossible to men who had not studied the highest parts of the theories of their subject; and that work may be taken as a type of what is likely to happen in the future. Even in the smaller industries, where everything depends on utilising waste products, the highest theoretical knowledge is required, as is clearly shown in the case of those belonging to chemistry, a great many of which are now taken out of British hands, and monopolised by Germans, simply because they have brought to bear on them all available science, with the result that what were formerly considered waste materials now yield all the pro-fits of the manufacture. I doubt much if a great deal of the education given in this country is likely to lead to such results, as it is too much directed to enable the students to pass some University, Science and Art Department, Board of Trade, or other examination, and rapidly evaporates after that has been undergone. I hold very strong opinions on this subject, and shall try to enforce them with all the energy which I am capable of exerting. There is no more saddening feature in the modern systems of education than the importance which is attached by School Boards, School and College Commit-tees and governing bodies, and by men who aim at being leaders of public opinion in educational matters, to the number of passes made at examinations ; which, as Professor Blackie put it the other day, attempt to measure the unmeasurable. Education, like every thing else, if it is really to progress, must submit to follow the con-ditions necessary to development. If there is any fact clearly brought out by scientific investigation, it is that increase of quantity is not necessarily improvement, but is apt to be the very reverse ; what is required is improved average individual quality. The effect of the system of examinations which has come to dominate education in every department has been to produce that improved average in a wrong way, by tending to bring all to a dull uniform level. This subject is so important, I should like to impress it on the minds of the members, by quoting the opinions of men who stand high in the educational world, but this would occupy too much time. I shall content myself by giving the opinion of the late Under Secretary for Foreign Affairs, himself a very distinguished Glasgow student. Professor Bryce, in an article some time ago, said " Neither examining nor the conferring of degrees is essential to a university, and as the latter, if divorced from the former, becomes delusive and demoralising, so even the former may become dangerous to both students and teachers. Examinations, moderately applied, to test the thoroughness of a teacher's method, the accuracy and industry of a learner, are so useful as to be almost indispensable. Applied for the purpose of awarding honours and prizes, they are still useful, being powerful stimulants, but are liable to serious abuse. When they grow to be a controlling influence in a place of study, prescribed to the seniors what and how they shall teach, to the juniors what and how they shall learn, they are mischievous, and do more harm in one way than good in another. They destroy the teacher's freedom. They pervert the learner's mind. They encourage, I will not say cramming, because a skilful examiner detects and defeats mere cramming, but over-teaching, a teaching which attempts to give more in a given time than the mind can digest and assimilate. They force a student to aim, not at knowing a subject, but at knowing what to say about it at short notice. They weaken the love of knowledge, and substitute for it the passion for success and distinction, perhaps for pecuniary rewards." It may be said that some of the present systems of payment necessitate teaching according to the prescribed codes and syllabuses. This is a very good reason for having these systems modified. In the meantime, however, do not let us sacrifice the education of the students for the sake of a few additional grants. Use the systems in so far as they do not interfere with our ideal of the kind of instruction which should be given ; but do not allow them to dominate that instruction ; and at the same time let us strive to have them improved.*

---

* In the *Glasgow Herald* of 21st February, there is a report of an address by Miss Flora C. Stevenson, of the Edinburgh School Board, in which I find the following remarks regarding payment by results in elementary schools, but they apply with even greater force to the higher schools and colleges. She said : " Turning to the kind of education we are providing for our children or rather the kind of instruction we are bound to provide, unfortunately this was a matter which was beyond the control of school managers and school teachers. She said unfortunately because, although she believed every one would agree that there were essential branches of instruction which all children must receive, she also believed it would be an enormous

The next question to consider is, Where should the professional
education of engineers be given ? Sir Lyon Playfair, in speaking of
the Scotch Universities, has said, " In Scotland professional training
and license to practice have always been considered essential duties
of the Universities.   In fact, the Scotch Universities derive their
whole strength from their contact with the people.   As Antæus
derived his powers from contact with the earth, so do the Scotch
Universities from their contact with the occupations of the people.
Sever their connection with them, and enforce upon them the
English University ideal that education should be given for itself
alone, and without connection with life work, and the Scotch
Universities would perish from inanition in twelve months.   Their
reason for existence is that they are in the midst of a poor population,
which must work for bread, and that it is possible, according to long
experience, to infuse liberal and scientific culture into the occupations
of a people.   This union of culture with technical training is the
earliest conception of the university constitution."   The Universities
have hitherto afforded a training for the older professions of the
Church, Law, and Medicine,* "but now we have come to a time when
there is again a great change in human knowledge.   We are come

---

advantage if our school teaching could be released from the narrowing limita-
tions of the requirements of a code and regulations.   Money was, they said,
the root of all evil.   It was indeed at the root of many of the evils of our
present educational system.   It seemed to her impossible for My Lords in
Council to draw up such a code in the best possible way, or for school
managers to administer it, or for school teachers to teach according to its
requirements, so long as the question of a money grant was one of the most
important considerations which influenced them. *She had never heard any
one, either statesman or educationist, defend this system of payment by
results on any ground but that of convenience, and for the reason that there
was no other ground on which it could be defended."* I would take the liberty
of making a small correction on Miss Stevenson's remarks.   Money is *not*
said to be the root of all evil, it is the *love of money* which is so characterised.
Money is a very good servant, but when either men or institutions become
dominated by the love of money, they are reduced to a state of slavery.   I
think, however, it is probable that Miss Stevenson may not be correctly
reported on this point.

* Professor Stuart's Introductory Address, University College, Dundee

to a time when the occupation of the mechanic, of the artisan, of the busy and varied life of our cities, has begun to have new light break on it, has gradually grown into being capable of being systematised. A science is found to underlie and throw light upon labour. The material part of human life calls for its scientific treatment, and is capable of it. This is the age of a new university movement, which in all directions is manifesting itself either by additions to existing Universities, or by the foundation of new ones. This is the time when again the force of a new light is making men unvisionary, and new 'faculties' are being founded, careless of the Pope's permission to teach, seeking, therefore, not necessarily any degree, but seeking the right to teach by acquiring the necessary knowledge; and institutions are arising whose object is, as 800 years ago, to give the people what they seek—namely, the scientific basis of the occupation which is to be theirs in life. The demand arises too from all classes of society, and it arises simply because now, for the first time, it is beginning to be possible. This demand is what is called at the present day the demand for technical instruction." The word technical is sometimes seized hold of by those who use words without considering the meanings which have come to be attached to them, as a pretext for saying that technical education should not be given in the Universities. I think it is unfortunate that so much prominence has been given to the term, as when used in the sense in which I am now using it, it has exactly the same meaning as scientific training, has when applied to medical men. They attend lectures, laboratories, and hospitals; the engineers should attend lectures, drawing offices, laboratories, and workshops or manufactories, and the methods adopted should be essentially the same in the two cases. If time permitted, I might quote from the evidence given before the Commissioners on Technical Education to show that a strong feeling exists on the Continent that the German Polytechnic Schools should be Faculties in the Universities, not only in spirit and constitution as they are at present, but also in organisation. As, however, the Secretary of State for Scotland recently admitted, in a speech which he made in Edinburgh, that the main object of the

18

Scotch Universities was professional education, I shall assume that as granted. I wish merely to claim for the newer professions the same privileges as have hitherto been accorded without question to the older.

But even supposing this point fully admitted, the battle may be only half won, as the following extracts from two introductory addresses given at the beginning of last session in the University of Edinburgh, one by the Professor of Hebrew and Oriental Languages, and the other by the Professor of Engineering, will show. Each wished to indicate the conditions which were necessary for the future progress of his department. One of the conclusions of the theological professor was that we should "allow a student to have very considerable freedom to follow the bent of his own inclinations, and to specialise his studies from the very first within the University. The greater the variety of courses there were open to him, each crowned with its suitable degree, the better. We have, perhaps, hitherto aimed too much at producing uniformity of result, instead of fostering special excellence in one department, or even giving it anything like a fair chance of manifesting itself. Let our examination standards by all means be high, but let there be different avenues to our degrees, or co-ordinate degrees of equal value, and don't let us insist on every one coming up to the same standard in every subject, irrespective of inclination or aptitude."

On the other hand, the engineering professor said, "As regards the teaching of technical subjects I am averse to anything like specialisation. During the College stage of instruction, I would rather hear nothing about either civil, mechanical, mining, or electrical engineering. It is best that at college the student should know no such distinction. They are but the several branches of one great tree, and are alike dependent for nourishment on the same general facts and principles. It is the roots and stem of this tree, not its branches, that it is the business of the young engineer to explore during the initial stage of his education. To him at this time it is better that the science of engineering—just as the science of medicine to the young medical student—should appear one and indivisible. Later on, when general principles have been mastered and some practical experience gained, he may, if he will, and the bent of his career requires it, become a specialist." And to confirm this opinion, he quotes that of Professor Huxley about medicine that, "Now-a-days it is happily recognised that medicine is one and indivisible, and that no one can properly practise one branch who is not familiar with, at any rate, the principles of all." And he further appeals to the future experience of the students by saying: "Many young men will, I doubt not, pass from these walls to seek employment in distant parts of the empire. To them a knowledge of the principles of *all* the branches of the profession will be of the first consequence; for it is more than probable they will there find themselves in the position of pioneers and men-of-all-work in the exercise of their calling, and be expected to undertake work of the most varied kinds. Men placed thus—and I have myself known them—who have had an exclusively civil or mechanical training, as the case might be, have not infrequently had cause to bitterly regret the one sidedness of their knowledge."

I am strongly disposed to side with the theological as distinguished from the engineering professor. As regards the requirements of engineering students the common-sense view of the matter is simply this. They cannot, as a rule, afford to spend more than three or four years at college or university, nor, perhaps, is it desirable that they should; if we attempt in that time to give them even the principles of *all* the branches of engineering, we must deal in the vaguest generalities, which the students have not the slightest idea of how to apply, and they pass through the course of study receiving neither education nor instruction, in the proper sense of these terms; and I know, from contact with many engineers, that a very strong feeling exists against a college course as at present arranged in many cases. If, by the principles of *all* the branches of engineering, is meant a good grounding in mathematics and physics, no one will be disposed to dispute the advisability of such an arrangement; but, if a smattering of half a dozen branches of the profession is meant, very few among practical engineers will be found to agree with it.

The school education of engineers should be such as is best fitted to make *men* of them in the proper sense of the word ; it should be liberal in the sense of training their minds to methods of independent thought, and of making the resources of literature and art available to them ; but at the same time it ought to lay the foundations of their future professional training, and the first half of their college or university career would be largely devoted to those subjects which are required in all the departments of engineering (*e.g.*, mathematics, physics, &c.), while the latter part should be specially devoted to the particular branch which they are likely to follow, and if they can take more than one of those special branches so much the better. In short, while trying to make their education as liberal as possible, they should endeavour to know at least *one* subject thoroughly. The argument quoted above about medicine being one and indivisible, does not apply with much force to engineering.   Medical men deal only with the human body, which must be considered as a whole, for the oculist who tried to cure the eyes without consulting the state of the stomach, or the aurist who treated the ears without taking into account the general state of the health of the patient, is not likely to do much good ; but engineering must be looked upon as a combination of many practically distinct professions (all, however, requiring the same kind of preliminary training), and it is absurd, for instance, to compel the future mechanical engineer to study all the details of tunnelling, or cement making, or railway construction, or the intricacies of hydraulic problems, so long as he has not had any opportunity of making himself acquainted with the theory of the steam engine, or of machines generally.   Not that the former subjects are not valuable in themselves, both as subjects of education and instruction, but the latter are equally so ; and if time cannot be found for both, then by all means drop the former ; and similar illustrations might be given from the other branches of the profession.

With regard to the training of those who go to the colonies or other different parts of the empire, no mere smattering of general principles is likely to stand them in good stead, unless they are at the same time thoroughly well up in the details of at least one department of the profession, and the obvious advice to such men would be to take more than one of the special branches we have mentioned. No doubt a very useful course of study might be arranged for some men who intend to proceed to the colonies, which would be a combination of civil and mechanical engineering (in the usual sense of these terms) and of agriculture. I understand that such a course as this is actually being given in the Crystal Palace School of Engineering; but such men could scarcely claim to be considered engineers, at least in the ordinary acceptation of the term. *

That engineering is not one and indivisible, it is sufficient to point to the Chair of Naval Architecture and Marine Engineering, recently instituted in Glasgow University. This being recognised, it ought to be followed to its logical conclusion. A system which grants a diploma in Civil Engineering and Mechanics, when not even the elements of mechanical engineering, in the sense in which that term is understood in a Continental or American college, are taught, and which allows a department of Naval Architecture and Marine Engineering to be constituted, without making arrangements for imparting instruction in the latter subject, and which does not even recognise the various other departments of engineering I have mentioned, cannot be considered logical. Surely a representative of this Institution in the University Court would at least have taken steps to show what ought to be done in these matters, and invited the support of the profession in endeavouring to place them on a proper basis.

I will now proceed to give some idea of what has been done in

* Since the above was written, I have seen a paragraph in the newspapers stating that the Professor of Engineering in Edinburgh University has organised classes in electrical engineering, to be conducted by his assistant. He has evidently changed his opinion about engineering being one and indivisible. I notice also that a special College for Colonial Engineers has been organised in England. In Glasgow our existing institutions ought to have arrangements elastic enough to meet the wants of such men, without requiring a special college.

other parts of the world to provide engineers with education suited to the demands of the times, and which might with comparative ease be carried out in Glasgow. No other city with which I am acquainted presents such a favourable field for the organisation of technical education, and yet in no one has so little progress been made, at least in the higher departments. I need not occupy your time with details as to what has been attempted in this country, as probably you are well acquainted with them. A glance will therefore be sufficient. London has of course taken the lead, having the vast resources of the City Companies to draw upon. The College and Central Institute of the City and Guilds of London are now fairly organised and in a position to do good work, while University College, thanks largely to Professor Kennedy, gives very complete courses in different departments of engineering. In addition to the ordinary classes in pure science in that college, there are classes for Applied Mathematics and Mechanics, for Civil Engineering (in the sense in which it has been defined above), for Mechanical Engineering, for Architecture, and for Applied Chemistry; and in each of these departments there are laboratories, drawing offices, and small workshops, the object of the latter not being to save the students the necessity for experience in actual workshops conducted on commercial principles, but simply as an aid to the class work, and for developing the mechanical skill of the students. The engineering laboratory is one of the most complete in existence. A description of it, and of the work done in it, was recently published in the Proceedings of the Institution of Civil Engineers, by Professor Kennedy. The department of applied chemistry is subdivided into various sections, which, in addition to the ordinary chemical industries, include metallurgy and the chemistry of agriculture. The Applied Science Faculty of King's College, and various special institutions either in London or its neighbourhood, take up different departments, so that instruction in the branches I have enumerated is fairly complete, although a better organisation might be favourable to economy. The provincial colleges in Manchester, Leeds, Sheffield, Dundee, and Birmingham are all ahead of Glasgow in the matter of appliances for teaching engineering, while that of Liverpool has a few weeks ago received a munificent endowment for its Chair of Engineering, and for the equipment of an engineering laboratory of the most complete kind.

If we turn to France, we find an elaborate system of special schools for civil engineers, mechanical engineers, for mining engineers and metallurgists, for chemists, and for agriculturists. In each of those classes of schools the instruction is of a very high standard, and the consequence is that the university, in so far as its teaching functions are concerned, has been reduced almost to the level of a popular lecture institution, and only retains importance on account of the powers invested in it of granting degrees to students who have studied in other schools.

In Germany and other continental countries most elaborate technical schools are to be found, but as a rule these do not compete so directly with the universities as in France, as they are generally situated in manufacturing districts at considerable distances from university towns, or where they are in the same town they are for all practical purposes faculties of the university. The methods of instruction are as scientific and the standards as high as, for instance, in medicine, and if it were not for a remnant of the prejudice attached to the word technical, I am convinced that they would be carried on under one organisation, an opinion which, as I have already remarked, is confirmed by the evidence presented to the Commissioners on Technical Education. It would open the eyes of those who are contented with the existing arrangements in Glasgow if they made a leisurely study of some of the chief continental colleges, when they saw how thoroughly the work was done in every department. From the *programm* of the Zurich Polytechnic School, a copy of which I received a few days ago, I find that during the present session there are upwards of 230 different courses of lectures. These continental schools give, in so far as the more theoretical part of the student's training is concerned, as nearly perfect a course as could well be arranged, and if they could spare an equal amount of time to the thoroughly practical part of their training before they are placed in positions of responsibility, no fault could be found with the arrangement, beyond saying that it was too long; but when we find young men of 23 or 24 years of age, whose only experience consists in what they learned at college, or in hurried visits to works, no wonder that the engineers of this country who have gained their knowledge in the rough and ready school of the workshop, look down upon their theoretical continental brethren. Some even express the opinion that the cry for improved technical education is altogether unnecessary, that in our extensive manufactories we have the best technical schools in the world, and that the polytechnics of France and Germany are only necessary because the industries in those countries are not so extensive or so well conducted as ours, and further that notwithstanding all that they had done in the way of improving technical education, we still keep the lead in all industries worthy of the name. There is sufficient apparent truth in these opinions to make them pass with those who only take a superficial view of things, but if they become universally current they would ultimately lead to most disastrous results.

During the past 30 years immense progress has been made on the Continent in every department of manufacturing industry, but our national advantages for many years kept the real effects of continental rivalry from being clearly discerned. These effects are now being felt, although the causes have not always been clearly recognised British engineering had its origin in the genius of a few men, and its rapid development was due to the native sagacity of the people and the abundance of our raw materials, chiefly of coal and iron. This in conjunction with the political conditions and geographical situation of the country enabled Britain to become the workshop of the world. The start which it thus obtained was in itself a great advantage, for the mechanical instincts of the country have been developed through several generations, a fact which should be carefully remembered when we compare such a country as Germany with Britain. There it may be said that it is only within the past 30 years that manufactures worthy of the name have been carried on, and this in conjunction with the fact that the Germans are much more given to speculation than action, makes their recent progress all the more wonderful. British engineers have been generally evolved from the hard although expensive school of experience, and the abundance and cheapness of the raw materials did not necessitate any great attention to economy. With Germans, however, the case was very different; their materials were scarce, and they were 50 years behind us in practical experience. They have now made up to us largely in practical experience, and they excel us in theoretical training, and it is only now that the race for supremacy in the industrial world can be said to be run under anything like equal conditions.

If we turn to America, we find that their technical schools are almost as complete as those of the Germans. Some of these are directly connected with the Universities, and are rapidly becoming models for the thoroughness of their arrangements and the complete-ness of their appliances. For instance, the School of Engineering connected with Cornell University, under the direction of Professor Thurston (a copy of the Calendar of which is to be found in the library of this Institution), is an example of what can be done when an enthusiastic man is backed up by public support. The Americans are too sharp business people to allow any money to be wasted in unnecessary institutions; but still, during the past dozen years or so, technical colleges, or the engineering departments of universities, have been more developed in America than in any other part of the world. The conditions under which business is carried on in that country are very similar to those which exist in our own, and the objections which are brought against French or German schools as inapplicable to Britain, cannot be urged against those of America.

In order to give some idea of the nature of the training which this latter nation considers necessary for its engineers, I will give a

---

* An excellent paper on the training of mechanical engineers, by Professor Woodward, of Washington University, St. Louis, is to be found in the last volume of the Transactions of the American Society of Mechanical Engineers, which, with its accompanying discussion, I recommend to the attention of the members of this Institution.

sketch of the course of the Massachusetts Institute of Technology, of which I happen to have more detailed information than about any other. The School of Industrial Science of this Institute provides an extended series of scientific and literary studies and of practical exercises, and these are so arranged as to afford a liberal and practical education in preparation for active pursuits, as well as a thorough training for most of the scientific professions. The preparatory courses of study include the physical, chemical, and natural sciences, pure and applied mathematics, drawing, English, French, German and other modern languages, history, political economy, and international and business law; and special courses of applied science have been arranged in the following departments :

1. Civil and Topographical Engineering.
2. Mechanical Engineering.
3. Mining Engineering.
4. Architecture.
5. Chemistry.
6. Electrical Engineering.
7a. Natural History.
7b. Biology—preparatory to Medical studies.
8. Physics.
9. General course (intended chiefly for science teachers).

The ordinary course of study in each of those departments extends over four years. The subjects of the first year are common to all the courses, and are such as are considered essential as preliminary training, and as a foundation for the more strictly professional studies. These are—

FIRST YEAR.

First Term.	*Second Term.*
Algebra.	Plane Trigonometry.
Solid Geometry.	General Chemistry.
General Chemistry.	Chemical Laboratory.
Chemical Laboratory.	Political History.
Rhetoric.	French.
English Composition.	Mechanical and Freehand Drawing.
French.	
Military Drill.	Military Drill.

At the end of the first year, the regular students select the course which they will pursue during the remaining three years, and their work becomes more specialised as it progresses. Even in the special courses the students are allowed in the later years considerable option in their lines of study, so that they may devote themselves to one or more chosen branches of the professional or scientific subjects. In some cases the selection of later options is positively determined by the earlier ones, owing to the requirement of certain subjects as preparation for others; in others, a wide choice is offered throughout all the years, the difference in this respect arising largely from the nature of the topics involved. We cannot consider the details of all the special courses, but the following syllabus for Mechanical Engineering will give some idea of their completeness:—

SECOND YEAR.

First Term.	*Second Term.*
Principles of Mechanism.	Mechanism of Mill Machinery.
Construction of Gear Teeth.	Mechanism of Shop Machinery.
Drawing.	Drawing.
Carpentry and Wood Turning (shop work).	Pattern Work (shop work)
Analytic Geometry.	Differential Calculus.
Descriptive Geometry,	Physics.
Physics.	Literature.
Political Economy.	German.
German.	

THIRD YEAR.

First Term.	*Second Term.*
Slide Valve, Link Motion.	Steam Engineering.
Thermodynamics.	Drawing, Design and Surveying.
Steam Engineering.	Mech. Engineering Laboratory.
Drawing, Design and Surveying.	Forging, Chipping, and Filing (shop work).
Forging (shop work).	Kinematics and Dynamics.
Integral Calculus.	Strength of Materials.
General Statics.	Physical Laboratory.
Physics, Lectures and Laboratory.	Literature.
German.	German.

FOURTH YEAR.

First Term.	*Second Term.*
Mechanical Engineering.	Hydraulic Engineering.
Hydraulics.	Mechanical Engineering Labora-
Machine Design.	tory.
Mech. Engineering Laboratory.	Engine Lathe Work (shop work).
Engine Lathe Work (shop work).	Strength and Stability of Struc-
Strength of Materials.	tures.
Metallurgy.	Theory of Elasticity.
Heating and Ventilation.	Constitutional History.

Options.	*Options.*
1. Marine Engineering.	1. Marine Engineering.
2. Locomotive Construction.	2. Locomotive Construction.
3. Mill Engineering.	3. Mill Engineering.

Lectures are given in the different departments, and in the drawing offices the students are taught to apply what they have learnt to actual designs of machines, and sufficient shopwork is given to enable them to use their tools, and to understand their construction. The mechanical engineering laboratory is very complete, and the objects to be accomplished by it are the following :—

1. To give to the students practice in such experimental work as they are liable to be called upon to perform in the practice of their profession, as boiler and engine tests, pump tests, calorimetric work, measurement of power, &c.

2. To give to the students practice in carrying on original investigations on mechanical engineering subjects, with such care and accuracy as to render the results of real value to the engineering community.

3. By publishing, from time to time, the results of such investigations, to add gradually to the common stock of knowledge.

The laboratory contains as a portion of its equipment,—

1. An eighty-horse-power Porter-Allen engine, by which power is also furnished to the new building and to the mining department.

2. A sixteen-horse-power Harris-Corliss engine, used almost

entirely for experimental purposes : this is furnished, in addition to its own automatic cut-off governor, with a throttle governor, so arranged that either can be used, the former being in addition so constructed that the speed of the engine can be varied at will.

The exhaust of each engine is connected with a surface condenser, and thence with a tank on scales, so that the water passing through the engines can be weighed.

3. Two surface condensers, one of which is arranged in sections, so that the condensing water can be made to traverse the length of the condenser once, twice, or three times, at the option of the experimenter.

4. Machinery for determining the tension required in a belt to enable it to carry a given power, at a given speed, with no more than a given amount of slip.

5. Two brakes so constructed that a given amount of work can be put at will on either engine, and in such manner that this work can be accurately measured.

6. A steam-pump so arranged as to enable the students to make pump tests, indicating both the steam and the water cylinder, weighing the exhaust steam, and also the water pumped.

7. A six-inch Swain turbine-wheel so arranged that it can be run under a head of 15 feet, and that experiments can be made on the power exerted, the efficiency, &c., under different gates.

8. Two calorimeters.

9. A dynamometer.

10. Cotton machinery as follows—viz., a card, a drawing-frame, a speeder, a fly-frame, a ring-frame, and a mule.

11. Apparatus for testing injectors.

12. A mercurial pressure column.

13. A mercurial vacuum column.

14. Apparatus for determining the quantity of steam issuing from a given orifice under a given difference of pressure.

15. A good supply of indicators, gauges, thermometers, anemometers, and other accessory apparatus.

16. Four horizontal tubular boilers.   Another boiler, a forty

horse-power Brown engine, a number of looms, and other apparatus
n the mechanical laboratories on Garrison Street, are available for
the purpose of experiment.

As examples of the work done in the laboratory, the following
experiments are enumerated:—Tests of the evaporative power of
boilers; tests of the effects of different cut-off, compression, back-
pressure, speed, &c., of engines under constant or variable loads;
calorimetric tests; dynamometric measurements; investigations of
the tension required in a belt to carry a given power, at a given
speed, with no more than a given amount of slip; experiments on
the efficiency of condensers under different conditions; on the
efficiency of a turbine, &c.

In conjunction with this laboratory is another which is called the
laboratory of applied mechanics—although personally I do not think
it advisable to draw a distinction between an engineering laboratory
and one for applied mechanics.

The object of this latter is to give to the students, as far as possible, the opportunity of becoming familiar, by actual test, with the
strength and elastic properties of the materials used in construction.

It is furnished with the following apparatus :—

1. An Olsen testing machine of 50,000 lbs. capacity, capable of
determining the tensile strength and elasticity of specimens not more
than two feet long, and the compressive strength of short specimens.

2. A testing machine of 50,000 lbs. capacity, capable of determining the transverse strength and stiffness of beams up to 25 feet in
length, as well as of many of the framing joints used in practice.

3. Machinery capable of determining the strength, twist, and
deflection of shafting when subjected to such combinations of torsional and transverse loads as occur in practice, and while running.

4. Machinery for making time-tests of the transverse strength and
deflection of full-size beams.

5. A machine for testing the tensile strength of mortars and
cements.

6. The accessory apparatus needed for measuring stretch, deflection, and twist.

The classes are divided into small sections when making tests with machines.

All the experiments are so chosen as to make the student better acquainted with the resisting properties of materials, many of them forming part of some original research. Those on transverse strength and stiffness have also determined certain constants for use in construction, which had not previously been determined from tests on full-size pieces.

Time will not allow me to enter into further details of this institution, but when I say that each department is as complete as that of which I have given an outline, I hope I have shown very good reasons why we should not be satisfied with the arrangements which at present exist in Glasgow. I may add, however, that for the current session there are 76 members of the teaching staff in the Massachusetts Institute of Technology, including professors, associate professors, assistant professors, instructors, assistants, and special lecturers, and that the fees are 200 dollars, or about £40, per year. Notwithstanding this comparatively high rate, the number of students enrolled at present is 636. In the last edition of the catalogue of the school a list of its graduates is given, and the positions they now occupy, from which it is evident that the Institute must have exercised very considerable influence on the industries of America.

I have selected this American Institution for special reference, because the conditions under which it was founded, and under which it lives, are more nearly similar to what we have in this country, than others with which I happen to be more intimately acquainted. I might, for instance, have referred to French and German schools, but I would have been told that the conditions in those countries are so different to what exist here, and being for the most part subsidised by their governments, they would teach us little. I could, however, have pointed out that the largest and most successful French technical school—the Ecole Centrale des Arts et Manufactures in Paris—has all along been a self-supporting institution and not only that, but it has saved out of its revenue sufficient to pay in great part for its fine new buildings. At present it has upwards of 600 students, and more than 100 applications from properly qualified candidates for admission are declined every year, simply from want of accommodation, and there are now in France nearly 4000 of its graduates, in charge of engineering works of all kinds. I might have pointed to the Imperial College of Engineering in Japan, which I had the honour of organising, and in which very complete courses of instruction were given, in the departments enumerated at page 130. Details of these, however, are to be found in the Calendars of the College, copies of which are in the library of this Institution.

We should, however, in organising engineering education in Glasgow, not imitate any existing institution, nor guide the instruction by a syllabus of subjects drawn out by men who have no connection with the management or teaching of the local schools and colleges. The Scotch Universities have hitherto been the professional schools of the people, and it is to them that engineers should look for the higher parts of their training. But before they can adequately meet the demands of the present day, they must enlarge their ideas about their constitution and arrangement, and not allow them to be limited by the walls of a single college. At the beginning of the calendar of Glasgow University, we are told that in its early days the University had no endowments, and was possessed of no property, except a University purse, into which certain small perquisites were paid. We are, however, further informed that notwithstanding the smallness of its means, it continued to discharge its important functions with great zeal and activity, and attracted a greater number of members than could well have been expected in that rude period of society. The University at that time was simply an organisation for teaching, and we must return to this strictly conservative conception of its functions, a conception which I hope will be realised before very long. When, therefore, I speak of the University undertaking the education of engineers, I do not mean simply the College at Gilmorehill, but, in addition, any other institution in the West of Scotland, in which instruction is given up to a required standard, and which conforms to the regulations laid down by the University Court. Of these institutions, the most important in so far as the education of engineers is concerned, will no doubt be the Glasgow and West of Scotland Technical College. In my opinion this college ought, in its higher departments, in conjunction with the existing engineering chairs in the University, and others which may be founded, to aim at becoming the Faculty of Engineering, or Applied Science in the University. It may take considerable time before it makes sufficient progress to enable it to assume that position, but from the very beginning it ought to try to deserve it, and the governing body should take care that no arrangements were made which prevented its development in the proper direction. With a few extensions, even at present, comparatively complete courses of instruction could be given in the departments enumerated at p. 130, as with our excellent opportunities for practical work there is not the same necessity for subdivision of the subjects as there is in Germany, where each would have three or four ordinary professors, besides numerous assistants of various kinds. In any case, public opinion on the subject should be guided by indicating what is required, and no doubt if the work is taken up with enthusiasm by a few influential men, the necessary means will be forthcoming. But do not let us over-estimate the value of money in connection with the work of a University. It is, of course, useful, but by itself it cannot make a University successful. It is possible even to imagine a University with too much money becoming drowsy, as Oxford and Cambridge did, until a very few years ago, but it can never have about it too many men endued with the spirit of research, or with the wish to teach what is known of their subject, for efficient teaching is quite as necessary as research. We are told by historians that in the early ages of civilisation, all great movements were the work of volunteers. Surely with our boasted nineteenth century civilisation, a great amount of useful volunteer work might be done, if only the Universities led the way. We find men of wealth and leisure devoting themselves to missionary work of all kinds, for the improvement of their less fortunate fellow men, because they have seen the folly of mammon worship, or its accompanying fame worship.

I shall now sketch as briefly as possible the kind of training I consider suitable for young men who propose to enter the profession of engineering, and in order to make my statements more definite I will suppose that mechanical engineering is the department they have selected, although my proposals will apply with slight changes to the other departments. The problem is : Given 21 or 22 years as the maximum age to which the ordinary preparatory training of engineering students can be extended—and in this country I do not think it either advisable or possible to prolong it beyond that— what is the best use to which the time at their disposal may be put in order that they may be fitted to undertake the duties which usually fall to an engineer ? The ordinary school education must, in the first place, be very much improved, both as regards the subjects taught and the manner of teaching them. The first necessity for engineering, as for other professions, is of course that the students should have a good elementary English education. This should be continued to about the fourteenth year of their age without any great attempts at specialisation. At that age they should enter a secondary technical school, such as, for instance, Allan Glen's Institution. The ordinary secondary schools might, at least for some departments, give this secondary technical or scientific training. If the students intended to go direct to works from the school they might remain there till they were 17 years of age, but if they proposed to enter the college or university course they might leave at 16. By the time they are 16 or 17 they ought to be fairly well prepared in the ordinary branches, with the addition of elementary mathematics, physics, and drawing, and their simpler applications to the special department they mean to follow, and if possible have at least a working knowledge of one or two modern languages. Book learning should have a very subsidiary part in the course, the instruction in the different departments being made real and thorough by means of laboratories, museums, small workshops, and drawing offices, which would give the students the opportunity of making themselves acquainted with things, instead of simply listening to lectures about things. The practice of drawing ought to be encouraged, not by copying from badly arranged sheets, but by making it a real educative as well as useful exercise. The ability to draw is almost of as much importance as the ability to write, and the study of drawing should receive a proportionate attention in schools. The great mistake that is made in teaching drawing is looking upon it too much as a branch by itself, instead of making it the graphic expression of other branches, and an introduction to the actual drawings required in practice. Problems in arithmetic, algebra, and physics are often capable of graphic illustration, while the furniture of the schoolroom, and the school itself, could afford examples of the drawings used in ordinary construction, the easier parts of which could be taken to the school workshop and actually put into practice. The chief object of this workshop, however, should be to give opportunities for illustrating and applying what has been taught in the classroom.

It must not be forgotten in all the attempts at the improvement of education, that the first requisite of good living, of success in engineering, or in any other profession, is to be a good healthy animal, and that to be a nation of good healthy animals is the first condition of national prosperity. I would rather have a boy of 17 strong and comparatively ignorant, if his ignorance was free from vice, than puny and learned, if his weakness arose from overstudy. In those days of competitive examinations the objects of all education, or even of instruction, are very often forgotten. These are supposed to be the bettering of human condition and the acquisition of knowlege. As a matter of fact the sole object of many teachers and students is the passing of some examinations, after which the subjects are indefinitely shelved in most cases. This system is altogether wrong, intellectually, morally, and physically, and destroys the freshness and vigour which should be stored up for the hard struggle in life, and the student is apt to degenerate into a mere bookworm, wholly awanting in that manliness of thought and action which is the chief characteristic of a successful engineer.

On entering the college or university at the age of 16 or 17, the students ought to be able to produce evidence of proficiency in the preliminary subjects I have mentioned, in the shape of certificates from the schools they have attended, for the multiplication of examinations ought to be avoided, and only when certificates were not forthcoming, or when the students wished an independent examination should they be subjected to such. Courses of study for each of the departments I have enumerated at page 130, should be organised, and made as complete as the time will allow. These courses should extend over, at least, three years, in the first of which the subjects would be nearly the same for all classes of students, and would afford a foundation for the special applications to be given further on; while the second and third years would be devoted almost entirely to the special department they had selected. If time allowed them to take an additional year of study the elements of one or two other departments might be taken up; for while I wish to insist on each of the students knowing one branch thoroughly, it is very advisable that they should be acquainted with other branches. For instance, a knowledge of, at least, the elements of the theory and practice of mechanical engineering is of the greatest use to all classes of engineers, and when the time can be spared they ought to spend a year or two in a mechanical workshop. The ordinary civil engineer employs machines and engines of all kinds in the construction of his works. Shipbuilding is as much a department of mechanical engineering as locomotive building, only considerably more complicated. Electric engineering is simply a part of mechanical engineering with a special knowledge of the applications of electricity superadded. Chemical engineering, metallurgy, and mining engineering all require a knowledge of the use of machines and tools. It would, therefore, be highly desirable if all the students had a good training in the theory and practice of mechanical engineering before they took up their special department. The great majority of students can, however, spare neither the time nor the money for such a lengthened course, and, in my opinion, it is the duty of the

University to make arrangements for utilising the time which they can spare, not only for the purpose of giving them as liberal an education as possible, but also such a training as will be useful to them in their profession. This three years' course should not, however, complete the theoretical training of the students. After they have entered the workshop or drawing-office they ought to attend lectures in the evenings in the special department they had selected; and I trust that some of our leading engineers and manufacturers may be induced to give occasional lectures on subjects connected with their work, and that they will receive a hearty welcome from the University authorities, while others, having made enough money to enable them to live in comfort, will devote themselves to the advancement of the welfare of the rising generation, by engaging in work, either directly or indirectly, connected with the University. I have no doubt this proposal will be received by some with scorn, as altogether too ideal to be realised. That, however, will not discourage me. If lofty ideals are not to be found within a University, where, I ask, can we expect them to be found ?

I need not enter into details regarding the nature of the instruction which ought to be given in the various departments, as what I have said about the Massachusetts Institute of Technology gives a fair idea of these. I may, however, quote a few sentences from a paper I read some time ago before another society * which meets in these rooms :—

"The instruction in the special subjects should not be confined to mere lecturing, which, to a great many students, is very unprofitable. Lectures are useful if they teach the students to think for themselves—if they stimulate to exertion, and give advice which springs from personal experience—if they guard against wrong methods, and give information as to the latest results on any branch of the subject; but it is a waste of time, both on the part of the student and the teacher, if they are confined to a restatement of principles

---

* Proc. Phil. Soc. Glasgow, November, 1883.

which can be studied in print with greater advantage than from lectures. The teacher ought to use a text-book, and occasionally give a running commentary on those parts which seem to require it: but the greater portion of his time in the class-room should be employed in giving illustrations, and the results of practical experience and of original investigations. The most important part of the work, however, should be carried on in the drawing offices and laboratories. In the former, mere copying of drawings should be prohibited, and everything done from sketches taken by the students, from models or actual machines, or from data supplied by the teachers, all the details of the calculations, however, being made by the students. In the laboratories, experiments should be made to illustrate dynamical principles, to ascertain the properties of materials, and to test the efficiency of engines and other machines; while the electric engineers, the chemists, and metallurgists should have departments of their own, with all the necessary appliances; and the naval architects a tank for experiments with ships' models. To the various departments there ought to be attached small workshops, with a selection of hand and machine tools, not for the purpose of teaching the students the practical part of their profession, but simply to give them an elementary acquaintance with tools, and facilities for preparing their laboratory experiments, and opportunities for making such parts of apparatus as they may themselves have designed. Then there ought to be museums in each department, or a convenient central museum, where the students could spend some of their leisure time in examining well-arranged specimens of raw materials, of products in various stuffs, and of all kinds of machines and engineering appliances. I need not enumerate further at present, as I have already mentioned more than is likely to be applied for some years to come; but the system I recommend may be briefly summed up by saying that the professors should endeavour to make their instruction real, by acquainting the students with things themselves, instead of merely giving them information about things.

"The examinations, especially those near the end of the course, should be of such a nature as to test what the students know about their subjects, and not, as is very often the case at present, give prominence to the best crammers and the most expert writers. They ought to be placed under very nearly the same conditions as they would be in actual practice—that is, supplied with the necessary data and a few books of reference, and be prepared to show that they are able to apply their knowledge to the solution of practical problems."

With regard to the practical training of the students, I may further quote:—

"Different plans have been proposed for obtaining a knowledge of practice. The most usual recommendation is that the students complete their College course before beginning their apprenticeship, but in my opinion this is a mistake, as it is of the utmost importance that they begin their practical work as early as possible, as it gives them a great advantage in their studies when they have had some experience in the workshops. Moreover, young men of 20 or 21 are not likely to take to what is often considerable (but very necessary) drudgery, as those who are three or four years younger. To get over the difficulty it has been suggested that the apprenticeship should be served before entering College; but on the other hand, I am afraid that few would have patience to begin a lengthened course of study when they were 20 or 21. In Japan I got over these two difficuties by keeping the students at College in the winter and spring months, and sending them out to works during the remainder of the year, and, in addition, requiring them to serve two years under properly qualified engineers, and I think this is the system which ought to be carried out in Glasgow. It may be said that engineers would object to receive apprentices if their time was broken in the way I have suggested, but I think that there would be no difficulty in obtaining employment in workshops if the students went in as real working apprentices, and were paid the market value of their work, and made to understand that their business was to work."

Such a system would suit our Scotch University arrangements, with

their long summer recess, and would allow the practical experience of
the students to augment with their theoretical training. When they
had completed the course at the University, they ought to serve at
least for three years in practical work, attending in the evenings such
lectures as I have already mentioned. By the time they were about
22 years of age they would thus have obtained a training which
combined the good features of the Continental and the British
systems, and I think it would be found that we would soon have a
class of men who were able to take advantage of whatever opportunities they might have for advancing the work of which they were
in charge, and to more than hold their own with those trained in
Continental schools.

Although the training of foremen and the higher class of workmen, strictly speaking, falls beyond the scope of this paper, still it
would be incomplete if I did not say a few words concerning it, for
I have no doubt many of the best of those would strive to attend
the University courses, although at a more mature age than those
whose parents could afford to send them there direct from school.
Leaving school, as they generally would do, at about 13 or 14 years
of age, they would enter workshops, and those who were inclined to
improve their positions would attend evening classes in pure and
applied science. The great majority would find all they require in
these evening classes, but the most distinguished of them would
pass to the College or University day classes, and proceed to a
degree. This they should be allowed to do after about half the
usual attendance at these classes, as their practical experience in
the workshops, and their diligence in the evening classes, would in
that time enable them to surpass the standard attained by those
who had not this experience ; and I have little hesitation in saying
that it would be found that the best engineers were those who went
through this course, as they would have more self-reliance, energy,
and determination than those who attended the University classes
in great part to please their parents, but whose hearts were not in
their work. I would strongly deprecate making the passage from
the workshops to the University too easy, as those who were not

able to overcome considerable difficulties by their own perseverance, industry, and ability had better remain in the workshops. The tendency for young men to go in for some occupation in which they do not require to soil their fingers, is becoming far too common in all departments of life, and should certainly not be fostered by unfitting workmen for manual labour, when their abilities are not sufficient to enable them to rise to the higher departments of the engineering profession.

In both the day and evening classes much more attention should be paid to modern languages than is the case at present. The University of Glasgow has recently taken a step in the right direction, in its B.Sc. examinations, by recognising the importance of a knowledge of French and German, and this should be extended to every department of commercial and industrial education. Not only are these languages necessary to the engineer, in order that he may keep himself informed regarding the recent advances in the science and art of his profession, but they are now essential for ordinary business purposes. At the present time I wish to engage an engineer for a position in the East; but I find that the condition that he is expected to know French and German practically excludes all Scotch engineers.

For some special trades there ought to be trade schools, which would supplement the experience in the workshop, for while admitting that the workshop must be the main place where practical experience is to be gained, it cannot be denied that there are many subjects connected with workshop practice on which much useful information might be given in schools, and where also special practice could be obtained with greater advantage than in the workshop. In certain departments of engineering this is true, but it applies with greater force to those crafts or trades requiring high artistic or manipulative skill, such as carving in stone and wood, staining glass, dyeing, and weaving. For the most part, these schools would be conducted in the evenings, although in some cases it might be advantageous to have them also during the day.

21

I will not enter into further details connected with trade schools, and merely mention them to complete the scheme of education which I consider necessary to meet the wants of the present day, and which is represented in the annexed diagram.

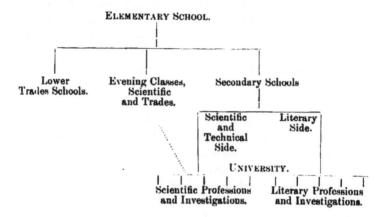

The dotted line from the evening classes indicating, as I have already mentioned, the possibility of passing from the evening to the University classes.

Before I conclude I should like to say a few words on another aspect of the subject. Technical education is only a small part of a much larger problem which we have to solve, and the danger is in the reaction against the older methods of education which have hitherto prevailed, the results will be more one-sided than those ever before produced. Although I have insisted on each man knowing one subject as thoroughly as he possibly can, there are no reasons why along with that, he should not have received at least the elements of a liberal education. Employers of labour too often speak of their workmen as so many *hands*, and rarely think or speak of them as having either heads or hearts, while managers of schools and colleges, if we may judge from their methods of instruction, devote their attention almost entirely to the heads, and forget that their students have human bodies which require care, or souls which are not satisfied with mere material wealth. Well may Ruskin exclaim that "the great cry that rises from all our manufacturing cities, fiercer than their furnace blast, is all in very deed for this, that we manufacture everything except men ; we blanch cotton, and strengthen steel, and refine sugar and shape pottery, but to brighten, to strengthen, to refine, or to form a single living spirit never enters into our estimate of advantages." Political economy has hitherto been almost entirely taken up with the consideration of the means of producing wealth, without regard either to its results or its distribution, but

> " Ill fares the land, to hastening ills a prey,
> Where wealth accumulates, and men decay."

The men who are successful in making great wealth, who are said to "get on," are those who have inherited good physical constitutions, and who have consequently great force and tenacity of character. The worry of a successful life (in the usual sense of that term) and the conventionalities of modern civilisation, however, soon tell on them, and their sons and grandsons are generally far from being their equals. On the other hand, the degeneration and waste of life among the working classes, especially in large cities, is so great, that it is said that they would become extinct after the third generation, were it not for the continual renewal from the already too much depopulated country districts. Concerning them the Royal Commission on the Housing of the Working Classes, reported, " that the state of things revealed by the evidence is so startling, so full of disgrace and danger to the country, that it should not in any case continue. The majority of the class on whom the wealth and prosperity of the country, and the safety of its institutions mainly depend, are living under conditions which must be regarded by all thoughtful readers of the evidence to be both shocking and intolerable."

A few weeks ago my esteemed friend, Dr Russell, the medical officer of this city, in a very thoughtful lecture on some of our social problems said, " The facts which have engaged us to-night, form the soil through which creep the roots of revolution and social anarchy, drawing to them the noisome elements on which they feed." And further, "He had dwelt strongly on the moral government of the universe, and we must remember that this includes the rich as well as the poor, the comfortable as well as the miserable, the respectable as well as the outcast. You may be selfish and have no thought of your poor brother, but you cannot get rid of him by not thinking of him. Brains hereditarily debased and hearts full of gall are dangerous to the state. You live in a country which especially needs all the hearts and brains of educated people in dealing with its difficulties. You live in a city which of all cities bears the heaviest burden of civic responsibility. In Glasgow settle the lees of an enormous wage-earning population. Every facility for communication, the spread of intelligence in health matters which moves first the best of the working classes outwards, everywhere tends to leave us with a central mass of people who cannot be trusted to live without supervision and guidance."

Here again people are apt to take a one-sided view of matters and think that in evincing an interest in the degraded poor, they have relieved themselves of further responsibility. The leaders of discontent, however, especially on the Continent, are not men with brains debased. On the contrary they are men whose brains have been whetted by the best attainable education, but whose hearts are embittered by personal injustice, or by viewing the social wrongs which everywhere prevail. Discontent among the lower classes may occasionally break out in aimless riot, but among the educated it may lead to an explosion which will shatter our social fabric to its foundations. Shutting our eyes to such matters is a cowardly way of proceeding, neglecting to make arrangements which will at least tend to alleviate them is criminal. Suggestions for educational and social reform have hitherto been altogether one-sided, and have neglected to take into account the complex character of human nature.

My reason for introducing these considerations at the present time was not for the purpose of discussing them fully (although

I consider any subject whatever which affects industry as a legitimate subject for discussion before this Institution), but simply to
show that I had not overlooked them, and that I mean to do what
I can to prevent them from being overlooked by others. In the
past, social errors have been mainly errors of principle, whereas
the majority of people content themselves with discussing errors
of detail, which, although involving waste, its amount is very
often small compared with that resulting from errors of principle.
In the present day, too much is expected from technical education, at least in its ordinarily accepted sense. By itself it will
rather tend to intensify some of the evils from which the world
is at present suffering, by increasing the relentless competition
which already everywhere exists. There are grave reasons for
doubting if our modern systems of education and of manufacture
have augmented the happiness of the world. Thinking men begin
to see that what passes as the highest education is often nothing
more than learned pedantry, and, in pursuing that too exclusively,
we miss altogether the objects of life, the development of our own
nature, and the good of our fellow creatures; and a strong suspicion
exists that scientific and industrial progress has outrun moral improvement, and has thus been one of the factors in producing the
present state of unrest and discontent prevailing all over the world.
The solution of our social problems is not to be brought about
by the single remedy of the enthusiast, not by education (either
technical or liberal), not by *laissez-faire*, land nationalisation, social-
ism, or communism. The thoughtful statesman or the social philo-
sopher will carefully weigh each suggested remedy, and find what
virtue, if any, is in it, and his measures will contain as much of each
in due proportion as he finds good. He will recognise that real
progress must be made simultaneously along all lines, material,
intellectual, and moral, and that to remain prosperous, nations
must possess the conditions which Herbert Spencer prescribes for
individual welfare—"A constant progress towards a higher degree
of skill, intelligence, and self-regulation—a better co-ordination of
actions, a more complete life."

The discussion of this paper took place on the 6th April, 1887, when

Mr DYER said, before the discussion on the paper began, he would like to make a few remarks as to what had been suggested by another association with which he was connected for the re-organisation of the Scottish Universities, and especially for the amendment of the constitution of the University Courts. For some months past a Committee of the University Council Association of Glasgow had been considering these subjects, and it had proposed that the following should be *ex officio* members of the University Court, viz. :—the Principal and the Rector, the Lord Provost of Glasgow, the Chairman of the School Board of Glasgow, the Dean of the Faculty of Procurators of Glasgow, and the President of the Faculty of Physicians and Surgeons ; and further that the Dean of Faculties and three assessors should be elected by the Senatus Academicus, and five assessors should be elected from among the members of the General Council, by the General Council, as members of the same Court, which thus would number fifteen members in all. The following were the more important duties which it was suggested should be performed by this re-organised Court :—" The whole administration and management of the revenue and property of the University, and the college or colleges thereof, including funds mortified for bursaries and other purposes, shall, from and after the commencement of the Act, be transferred from the Senatus Academicus and be vested in the University Court. The University Court shall have the sole power to appoint professors whose chairs are, or may come to be in the patronage of the University ; to appoint examiners and intramural lecturers ; and to grant recognition to extramural teachers, and after the expiration of the powers of the Commissioners under the Act, it shall have the sole power to enact, after consultation with the Senatus Academicus, regulations for graduation and conditions for the tenure of fellowships, scholarships, bursaries, and other foundations. The University Court shall have power to take proceedings against a principal or professor, without the necessity of any one not a member of the Court appearing as prosecutor, it shall have power to define the nature and limits of a professor's duties under his commission, subject to appeal to the Universities Committee, and lastly it shall have power to establish new professorships, with the consent of the Chancellor and the approval of Her Majesty in Council." These were very important duties, and it seemed to him (Mr Dyer) that in order that the interests of engineering education may be properly attended to, it was highly desirable that this Institution should be represented on the University Court. At the last meeting of the Committee of the University Council Association, he had mentioned Mr Weir's proposal. It was favourably received by the members, but it was thought desirable that this Institution as a body should in the first place express its opinion regarding it. He trusted that the motion about to be made by Mr Weir would receive their cordial support.

Mr T. D. WEIR, B.Sc., C.E., said he rose to propose a motion which Mr Dyer had already described, and to which he had made some reference in the opening of his paper. Mr Dyer's account of the proposal which had been laid before the Secretary of State for Scotland, for the reconstruction of the University Court, was of greater authority than anything which he (Mr Weir) could have said, seeing that Mr Dyer was a member of the Committee of the University Council Association of Glasgow. His (Mr Weir's) information was based solely on the reports which appeared in the newspapers, while Mr Dyer, as a member of that Association, and an active member of the Committee, made his statement with the authority of the Committee itself. He did not propose in introducing the motion to the meeting to go over the paper that Mr Dyer had read. He might refer to it incidentally in support of his motion, but he did not propose to discuss the subject of the education of engineers in its various details. Mr Dyer's paper was a very wide one, and the more practical details, he had no doubt, would receive attention from other members. What he would like to draw attention to

was the acknowledged defects in the present system of engineering education which co-existed with a regular University course,
and he rather thought the blame must be laid to some extent
on this Institution. The Chair of Engineering dated back now
some 46 or 47 years, and there had always been a fair amount of
connection between it and the Institution. Their first President
was the holder of the Chair of Engineering of that day, Dr Mac-
quorn Rankine, and the present incumbent of the chair was well
known to the members of this Institution, and had also filled their
presidential chair. But the connection had been rather one-sided.
No doubt individual members of that Institution had shown their
appreciation of the benefits of a University course, and many of
their members—such as Charles Randolph, John Elder, and William
Pearce—were benefactors of the University. Unfortunately in
Scotland as compared with England, and more emphatically in
Great Britain as compared with Continental countries, in educational
matters they could not look for much governmental aid : they must
look to a great extent to the purses of the locality. The Glasgow
University had benefited through many members of this Institution,
but those benefactions had been given in a hap-hazard way. There
had been no official recognition of the wants of the Institution on
the part of the University any more than there had been any recognition on the part of the Institution of the wants of the University.
What they wanted was a closer connection between this Institution
and the University, and he could see no other way of obtaining this
than by direct representation in the University Court. There was
a precedent for them as an Institution having some control of engineering education. When the Mechanics' Institution was formed
into the College of Science and Arts, they as an Institution were
invited to send a member to its governing body, which they had
done. Now that College had been merged in the Glasgow and
West of Scotland Technical College, and the Council of this
Institution possessed and had exercised the right to elect a member
of the new governing body. In that way they had an opportunity
of putting forward their views in the education of engineers and

others in the initial stages; and the object he had in view, and which Mr Dyer had in view also, and which they ought all to have in view, was that their practical influence should extend to the higher branches—to the University training of engineers. He thought he had said enough to introduce the motion, and he now begged to move as follows:—"That in prospect of immediate legislation affecting the Scottish Universities, the Council be authorised to take action with a view to securing the direct representation of this Institution on the governing body of Glasgow University."

Mr P. S. HYSLOP, C.E., said he had very much pleasure in seconding the motion of his friend, Mr Weir, whose remarks, together with Mr Dyer's paper "On the Education of Engineers," had rendered a substantial service to the profession. They had brought before them a subject the discussion of which would, he trusted, result in such action being taken by this Institution as would tend to the immediate removal of an anomaly, and secure for the future that combination of theoretical with practical training which experience showed to be absolutely essential for young engineers. That the present condition of University education for engineers was anomalous must be patent to every one who was at all conversant with the subject, for who had not heard it remarked amongst practical engineers, civil or mechanical, that the University certificate was hardly worth the paper on which it was written. Many sad cases of disappointment had occurred through young men being led to place too great faith in purely University training, and it too frequently happened that a young man of 20 or 21 years of age, after working hard for several years to obtain the certificate, found, on presenting himself at an engineer's office, that his diploma, of which he was so proud, was practically worthless, that he himself was considered to be not only useless but rather a nuisance, and that he was told if he wished to be an engineer he must go and serve his apprenticeship. Such a case came under his own notice very recently, and similar cases must have occurred in the experience of most of the gentlemen present. He was far from under-rating the

22

value of pure theory, without which no real progress could be made in applied science, and nowhere had he seen its advantages more clearly pointed out than by Professor Rankine in the preface to his "Applied Mechanics." But Rankine as clearly pointed out the necessity of practical knowledge, and unfortunately the tendency of late years had been to leave out the practical part. Such a state of matters was surely not creditable to this country, which had hitherto occupied such a prominent position as a field for training men to do useful work in the world, and while that position was being maintained in other branches of knowledge it could not be denied that in the training of engineers other countries were going far ahead of this country. The abstract principle of combining practical with theoretical training was not a new idea. Ever since the medical profession existed as a profession such a combination had been practised, while no law student would ever dream of attending so many sessions at college, passing his examinations, and immediately fixing up his brass plate. Now, as compared with the other professions, engineering, in the wide sense of the term, was by far the most important agent in the advancement of civilization; indeed he went further and asserted that it was of more importance to the general welfare of mankind than all the other professions put together. Why, therefore, did not the thorough and efficient training of engineers receive attention in University work commensurate with the importance of the subject? It might be objected that the general welfare of mankind was not the immediate object of University work, but what, might he ask, would this island of Great Britain be, what would this city of Glasgow be, but for its industrial enterprises, all of which were dependent on engineering science for their initiation and development? Why then does not this great subject receive the attention it deserves? He believed it was simply because the engineering profession as a body had not insisted on its doing so, and he was of opinion that a grave responsibility would rest on them as an Institution if they allowed this favourable opportunity to slip without securing a powerful and effective representainto of the profession in the governing body of Glasgow University.

But in order that that representation may be effective it would be necessary that they should determine clearly the particular functions that should appertain to the University in the young engineer's course of training. Mr Dyer had sketched very clearly the course pursued by some foreign Universities, but it seemed to him that however much the system of University workshops might be elaborated they could never give the youth that facility and quickness for getting through actual work that was every year becoming more and more necessary to enable him to earn a livelihood on the completion of his training, and which he believed could only be secured by their own system of apprenticeship. It might be taken for granted that nowadays the possession of a University degree by a young engineer was in every way desirable, as it was or should be a certificate of his proficiency in his profession; but it was not more actually necessary for him to attend a specified number of lectures at the University, as was now prescribed, in order to qualify himself for obtaining that certificate, than it was actually necessary for a young barrister, in order to qualify himself in law, to eat so many dinners at the Temple, as was now prescribed for him by English law. Students of the other professions had their time divided between work and study in a manner for which they had every convenience, but with engineering students the case was wholly different, their spheres of work being scattered over large areas the distances to be travelled rendering it impossible for them to give due attention to their studies in the University and their work in the shops, or in the case of civil engineers in the field, at the same time. The University examination being assumed to be a test of knowledge it appeared to him to be a matter of the very smallest importance how and where the knowledge was obtained, whether in the lecture halls of Universities, in the class rooms of more popular institutions, by employing private tutors, or by private study. That seemed to be Mr Dyer's opinion, also, from the sketch he gave on page 160 of his paper. So long as a youth possessed the requisite knowledge he should be entitled, on payment of certain fees, to demand examination by the University authorities and the University degree if he passed the examinations successfully. In order that these examinations might be a test of working capacity, as well as book-learning, they should include the performance of actual work, and instead of being crowded into an hour or two should extend over a week, two weeks, ten weeks, if necessary, in order that the test applied might be of a thorough and a practical nature. If possible, it should further be established as an absolute rule that no one should be permitted to call himself an engineer until he had obtained such a degree, so that the status of the holders should be conserved. Doubtless this would cause difficulties with regard to men in actual practice, but he did not believe that such difficulties were insuperable ; on the contrary, he was of opinion that they would easily be got over.

General FRANCIS A. WALKER, of the Massachusetts School of Technology, writes as follows : — I have received your letter of the 5th March, in which you do me the honour to ask me to offer some remarks with reference to Mr Henry Dyer's account of this School of Applied Science, in his very valuable paper presented to the Institution of Engineers and Shipbuilders in Scotland. Agreeably to your suggestion, I shall request the gentlemen having charge of our several departments to prepare memoranda regarding the courses of instruction, apparatus, lines of experiment and research, &c. Unfortunately the time that will elapse before the next meeting of your Institution will not permit the preparation of these notes in season for the discussion of Mr Dyer's paper ; but what we may send will perhaps serve the purposes of the gentlemen who may be charged with editing the proceedings. At present I will only say that this school, incorporated in 1861, was opened to students in 1865, and graduated its first class in 1868. The several departments of the school are correctly enumerated by Mr Dyer. The land on which our two main buildings are erected—each about 90 × 160 feet on the ground, four stories and basement, in height—was conceded by the Commonwealth of Massachusetts. This land could, at the date of that concession, have been bought for about 135,000 dollars. Its present value is several times that sum ; but the fee is not in this school, but in the State, only the right of perpetual occupancy having been granted us. The school also receives the income of a fund of 115,000 dollars, the proceeds of the sale of public lands granted by the Government of the United States. About one million of dollars has, at various dates from 1863 to the present time, been given or bequeathed to the school. The great portion of this was necessarily put into buildings, plant, and apparatus. A small portion has necessarily gone to current expenses. About 400,000 has been invested in income bearing securities. Our tuition fees are, by the logic of our situation, unfortunately high, 200 dollars a year. The receipts from this source the current year are about 125,000 dollars. We have also some small receipts from rents. Our expenses are about 160,000 dollars a year, of which, perhaps, 116,000 dollars go in salaries, and 24,000 dollars in laboratory supplies. In addition to the two main buildings spoken of we have erected, on lands bought by the school, mechanic shops for instruction in carpentry, wood-turning, engineering and foundry work, and machine work on the metals, covering about 25,000 square feet. I send by this mail a copy of our annual catalogue and of the last President's Report, which will be found to contain much of the information needed to make up such an account for each of our several departments, as Mr Dyer has so felicitously drawn up for the department of mechanical engineering.

Mr ALBERT F. HALL, Charlestown, Boston, Massachusetts, writes: —Mr Dyer's paper on the Education of Engineers was received to-day, and has been read with much interest. I have the honour of being a graduate of the Massachusetts Institute of Technology, and can assure you that you can hardly form an idea of the extent of work done there. It is hoped that a very full account of the Institute and its work in detail, as written by my friend, Professor Lauza, will some day be printed. It has no equal in this country, and is very *thorough*. I have visited many of the schools in Germany, in Stockholm, and at Belfast, Ireland, also the schools in Paris, but it seems to me they appear much better on paper than in reality. The tests at the Institute are very thorough, both in the

manipulations and calculations. They have boiler tests lasting a
week, day and night, changing watches.

Mr A. W. THOMSON, B.Sc., said Mr Weir's motion, " That steps
be taken to obtain direct representation of this Institution in the
University Court," if passed by the Institution, would not, he was
afraid, commend itself to the University Court; but failing this direct
representation, the Institution would still have the professors who
occupied the technical chairs amongst its officers or members. For
some years he had been associated with Prof. Alexander on the staff
of the Imperial College of Engineering, Tokio, Japan, of which Mr
Dyer was principal; he had therefore had a good opportunity of knowing how the work of that college was carried on, and could bear testimony to the excellent results obtained by the general arrangements
instituted by Mr Dyer, and carried out in detail by his colleagues.
He thought Mr Dyer should have given extracts from the Japanese
Engineering College Calendar, showing the arrangements of the
various departments, for comparison with those of the Massachusetts
Institute of Technology. Mr Dyer thought that Glasgow should
"not imitate any existing institution;" in his opinion they ought to
imitate these so far as their various arrangements commended themselves to gentlemen who were teaching in their midst, and who were
interested deeply in their educational affairs. Many teachers and
other authorities on educational matters decried examinations, and
seemed to think they were unnecessary evils; in an extract from a
paper by Professor Bryce, and quoted by Mr Dyer, it was stated that
examinations were almost indispensable; he felt certain that examinations, for the purposes for which that writer mentioned, were not
almost but altogether indispensable; at the same time he felt certain
they were not always fair tests. He was inclined to ask people who
wrote thus—What would you propose instead? At a college or similar
institution, examinations in any subject should be frequent, and, if
possible, be arranged in two sets—those in which books of reference
might be freely used, and those in which the pupil's memory and
judgment must alone serve him. Such a method was adopted with
considerable success in the Engineering College, Tokio. He feared

that on pp. 144 and 145 Mr Dyer expected too much of his boy of 16 or 17; he thought drawing, until the student could *draw well*, should be considered "as a branch in itself;" and there was no reason why he should be asked to copy "badly arranged sheets." The idea of leading engineers and others giving occasional lectures on subjects connected with their work appeared to him to be a very happy one, and he should be glad to see it acted upon; such lectures to be useful should be of a particular, rather than of a general nature. Principal Caird, in addressing the students on one occasion said, that "The University did not wish to change a very good joiner into a very bad minister;" this same idea, as applied to the engineering profession, was well put in Mr Dyer's paper. He was sure that they all felt for the toiling millions who had to live in many cases—could they call it living?—in surroundings so miserable as those described by Mr Dyer.

Mr ROBERT DUNCAN (Whitefield) said he wished to congratulate the Institution on having in Professor Dyer a member who had not been afraid to break the ice covering a set of subjects of more importance to the whole of them than even the merits of steam jackets, or the latest departure in valve-gears. He particularly referred to Mr Dyer's closing remarks, pages 160 onwards, with nearly every word of which he most cordially agreed, and if any support from one who was an employer and not an outsider would help their effect, he felt it ought not to be withheld. It became clearer to him every year that permanent progress and success in their industries could only be founded on cordial human relations and recognition of mutual interests between employers and employed. Strict discipline, but strict justice, and the promotion of thought and exertion for one common end—the success of the business undertaken—would give enduring results that could only be imitated for a time by the driving factory system. No state of society based on injustice could last long, and they had recently seen a good deal of the explosive kind of termination that might be looked for to such a state. Let society get into the condition of a chemical mixture in unstable equilibrium, and a very small spark would produce explosion. There could be little

doubt that they had proceeded far on the way, both in this country
and others, certainly including the United States, to that state of
unstable equilibrium.   How this was to be stayed, and how a safe and
healthy state of society was to be produced, was now the question of
questions.   They might feel assured that if at all, it would be by a
growth, comparatively slow like all healthy growth, but sure, of
conditions and arrangements making life not merely tolerable but
interesting and ultimately beautiful to the great masses of the
workers.   Upon the seal of their Institution was seen the honoured
head of James Watt, and they were proud to be his countrymen,
and to think of the distribution of intelligence and ideas due to his
thought and work ; but let them also remember that they were the
countrymen of Robert Burns and Thomas Carlyle, representing the
poetry and the duties of human brotherhood.   These men in their
different ways had shown the same truth, that men had to recognise
themselves as banded together for common help, and that this was
only possible by loyal obedience to capable leaders.   He had been
looking with some interest into the reports of industrial co-operation,
and had seen what had not surprised him, that while societies for the
comparatively simple task of distribution of commodities for mutual
benefit had been exceedingly successful and were growing steadily,
that on the other hand societies for co-operative production had all
been failures.   The explanation of these latter failures to his mind
was, that there were not the necessary conditions resulting in capable leadership and thorough discipline.   He had also seen without
surprise and with much pleasure that there were a few instances
(though not in this country) of successful—remarkably successful—
manufacturing establishments where the principle had been adopted
of recognised common interests of employers and employed, where
the latter had shares in the profits of the concern, according to
certain definite rules, taking into account the nature and the duration
of service.   These were the establishments which surprised outsiders
by remaining busy and prosperous while others were languishing.   He
would venture the prediction that this was the road they had to take,
with slow and cautious steps at first, to gradually reach the transfor-

mation of the present organisation of labour to safer and more enduring ones. The "windy gospel" that there ought to be no capitalists would find few believers, at any rate among our hard-headed race, when every man saw the practicability of becoming himself a capitalist, and stable member of society. Shipbuilders would say with much justice that this looked like a dream to them while the present extreme fluctuations of trade existed, with the consequent desultory connections of employers and employed. He thought, however, that the more labour was banded together over the world, and the more intelligent our working population became, the less would there be of this extreme irregularity. The two mutually-affecting actions would require to advance together, and thus customers and producers gradually unite. The race of individual capitalists, as seen specially in the United States, who made good trade or bad trade as they stretched out or held back their hands, would probably tend to extinction, and at any rate their influence would be swamped by the great sea of small capitals intelligently combined for the general welfare. It seemed to him that this subject, and other kindred ones, were very suitable for occasional discussion in this Institution, and would supply what had been greatly wanted, something to bring out the interests and experience of the great employers of labour they had amongst them. Their President, so prematurely lost to them, seemed one marked out to give a powerful helping hand to the work he had spoken of. They were thankful that much of Mr Denny's handiwork remained. Professor Dyer had referred to their position as Second City of the Empire. He did not think, from what he had seen in a good deal of moving about, that it would be very difficult for them to become the first as regarded commercial morality. With Clyde yards turning out good sound work, Clyde employers capable and honest men, Clyde workmen doing their utmost for the prosperity of all, they had no need to fear that the trade which was worth cultivating would long be diverted from them by other channels.

Mr ED. C. DE SEGUNDO, on the invitation of the Chairman, said he had read with much pleasure the paper of Mr Dyer, and had followed attentively the previous speakers. He would like to make a few remarks on the interdependence of theory and practice. Theory and practice might be considered to stand at either end of a chain whose several links were functionally connected. One of our best-known professors of engineering had expressed his opinion that practice *was* theory; he would rather say that practice was theory multiplied by a co-efficient which would include and allow for various imperfections in the reasoning, arising from an insufficient knowledge of the laws of nature. He should say that the value of scientific reasoning (he used the word in its broad sense) in any occupation would be recognised by all broad-minded men, and that he might have started with the assumption that theory was as essential to practice as practice was to theory. But some men were still of the opinion that an ounce of practice was worth a pound of theory. Now, this was absurd. A pound of sound, correct, impartial theory was worth a pound of practical experience. It was impossible to fulfil in practice all the conditions that were assumed—as a rule—in theoretical investigations, but that was no reason why theory should be despised. In scientific work the incompleteness of their knowledge of the forces at their disposal, and in practical work imperfect workmanship and friction (the laws of which were not yet fully understood) were responsible for the discrepancy which often arose between theoretical reasoning and practical results, and for the necessity (often) of the introduction of an empirical co-efficient by which results arrived at by purely theoretical investigations had to be multiplied in order to adapt them to the exigencies of practical experience. But to introduce an empirical co-efficient in any formula was, at the best, unsatisfactory; for our knowledge of the subject was not in any way increased. They had evaded the difficulty rather than overcome it. The knowledge thus obtained held only for that particular set of circumstances. The evil effect for these particular conditions was to some extent obviated, but the cause still remained, and would assert itself under other circumstances when perhaps they least expected it. Practical rules of thumb were like patterns.

Each was only useful in turning out one particular structure. It was not safe to trust to empirical formulæ, which were based upon the observation of effects, and not on a knowledge of the causes. Not that he wished for an instant to underrate the value of observation and experiment, for these were the only foundations for scientific speculation; but they should go farther than observe; they should inquire. To make progress they should work intelligently. To do this they must know *why* they worked in any prescribed manner, and the reason why was only learnt by close observation, careful experiment, and accurate reasoning. Unquestionably practice provided them, in the first instance, with the material for their investigations; but it was by a process of scientific argument that they deduced, from these crude facts, results which practice, from its very nature, would have been incapable of arriving at. Theory, if they liked, was speculative, but speculative argument was necessary for the suggestion of new experiments, the results of which, when carefully investigated, would form a valuable addition to their stock of knowledge. Would engineering science have developed as it had if the physicist, chemist, and mathematician had not made it an object of study? Did not the triple expansion engine, the Bessemer and other processes of making steel, and the more accurate calculations connected with the strength of materials, &c., bear witness to this? Undoubtedly, then, the theoretical and practical departments ought to receive equal attention in the training of those engineers who were to occupy positions in which they must think and act responsibly, and ought to work for the general advancement of engineering science. Now, compared with the means which do exist for becoming acquainted with the purely scientific problems connected with engineering, the means at the disposal of the young student of obtaining practical information were sadly insufficient. Evidently, for those who do not intend to spend their lives at the bench or the lathe, it could not be necessary to spend five, or even three, years at actual manual work in a workshop. Those who had studied at college, and had their minds well developed by reading, &c., would

find it dull and insufferable to spend that length of time in the
workshop, where information is obtained so slowly, owing to the
total absence of systematic teaching. Could not some arrangement
be made in which, by payment of a fee, students would have
the privilege of working under a competent journeyman, whose
duty it would be to instruct his pupils in the methods employed
in ordinary operations, and to make them acquainted with the uses
of mechanical appliances in general ? Under such, or similar
conditions, it would be possible to learn more in two years than
under the present system in four. What he considered was a great
defect in the present existing system of education for engineers,
was the fact that no means existed for distinguishing talented from
other men. In medical and other sciences they granted degrees ;
now, why should the system not be the same with regard to
engineers ? In Dublin they gave the degree of B.E., and the
London University, which was purely an examining body, were
moving in the way of modifying the B.Sc. degree into that of
B.E. ; but this examination, he understood, was not one that would
commend itself to practical men. He suggested that some practical
degree should be created, as a means of testing, not only the
theoretical, but the essentially practical knowledge possessed by
engineers.

Mr W. CARLISLE WALLACE hoped that the older members of the
Institution would take part in this discussion. He thought that
engineers themselves were largely to blame—he meant those who
had large engineering businesses—for the difficulty experienced in
the education of young engineers. Hitherto it had always been
considered the correct thing, that a young man, after having served
his five years apprenticeship, was then in a position to go out into
the world and call himself an engineer. Now cases had come under
his notice in which young men leaving school, say at the age of 16,
after getting a fairly good education, they went into the workshop,
and according to the present arrangement worked from 6 a.m. till
half-past five in the evening, the usua. working hours of the workmen. Very few of them, indeed, felt inclined to occupy their time

at theoretical work in the evening, and the result was that by the time they finished their apprenticeship they had lost what little theoretical information they had previously possessed. These were sons of gentlemen of good position, and expected to get into high places, and yet they were not even good tradesmen. They had had a year or so in the fitting shop, and a short time in the pattern shop, &c., but could do nothing well and knew nothing of the theory of engineering. What it appeared to him that was wanted was a system of technical education similar to what Mr Dyer advocated in his paper, or some arrangement whereby young men could attend classes in winter and work in the shops during the summer. There was always great difficulty in accomplishing this; for no doubt it broke up the routine of the shop, but it would be of great advantage to the young men themselves, for they would not have the drudgery of a long five years' apprenticeship, which he had no doubt came very hard upon some young men, as Mr Segundo said. They would go fresh from the work in the shops to the College course; and if some such arrangement could be made, he thought it would be well. He approved of a thoroughly exhaustive examination to test the knowledge they had gained, both practical and theoretical, and the granting of a degree or something that would show the proficiency in theoretical and practical work a young man had attained.

The CHAIRMAN (Mr Dundas) said that two of the former speakers had mentioned the desirability of having young engineers attending College in winter, and employed in the workshop in the summer. He thought there were great difficulties in carrying out that proposal; for under that system they might have their workshops denuded of apprentices one part of the year, and a great many crowding into them in the summer when they had neither places nor work for them. He thought the far better way would be to carry on their theoretical and practical education simultaneously, like the lawyers, who attended their classes in the morning and afternoon, giving the best part of the business day to practical work. He thought some arrangement like that might be made with far less interference to business. For his own part, he could not see how they were to take on assistants in the summer time, when trade was often dull, and keep their places open for them, or how employers could be expected to keep their work back in the winter waiting on those who were coming to them in the summer. He thought there might be no very great difficulty in getting employers to permit their apprentices, during their apprenticeship, to come to business say at 10 o'clock, or later, and to leave earlier in the afternoon than ordinary, and thus enable them to carry on both their theoretical and practical education simultaneously. He proposed that the discussion be now adjourned till next meeting, in order that Mr Hunter's paper might be read.

Mr JOHN THOMSON approved of the postponement of the discussion. He felt it of vast importance to hear the various views of the members on this subject, as he had the honour to represent the Institution in the governing body of the West of Scotland Technical College, who were now trying to weld the three bodies—the College of Science and Art, Anderson's College, and Allan Glen's Institution —into one concrete body, and he would be guided very much by what opinions might be enunciated there. He urged upon all the members to seriously consider this matter, and Mr Dyer's paper— which he considered of the very greatest value, and had come to them at a very opportune time. He thought they were indebted to Mr Dyer for that paper. He did not say that he agreed with Mr Dyer in every point; but his views were very valuable, and he thought it was the duty of all the members of this Institution to read and seriously study the paper, and come and give a full expression of their views at the next meeting. He held this to be of great importance, as it would be a valuable guide to him as their representative in expressing their views at the board of the Technical College.

Mr DYER begged to suggest that members might write out their views on the subject, and send them to the Secretary before the next meeting, as he was afraid they would have very little time at their disposal at that meeting for the discussion. He thought it was of vital importance that all who had studied the matter should express their opinions upon it.

The CHAIRMAN (Mr Dundas) thought it would be advisable for the members to carry out Mr Dyer's suggestion. There were many members who might thus put their views on record, and express their opinions on this matter, who might not be able to come to the meeting; and besides, it would save the time of the next meeting, when the discussion, he supposed, would have to be closed.

The suggestion was agreed to, and the discussion adjourned.

The discussion of this paper was resumed on the 10th of May, 1887.

Mr JAMES WILLIAMSON said, it would, he was sure, be admitted, that Mr Henry Dyer had rendered good service in placing before the Institution his opinions and experience upon the subject of the "Education of Engineers." That this paper had been drawn forth at a most opportune time and season would be equally conceded by all, whatever their opinions might be upon his manner and method of treating the subject. Mr Weir's motion now before them for a direct representation of this Institution in the University Court would, he trusted, lead to a very full and exhaustive discussion upon this important subject, that might result in some practical and decisive steps being taken to secure direct representation, not only for "engineering," but for "shipbuilding," and *all classes* of engineers, including mechanical, civil, mining, and military in the University Court, and also on the governing body of the Glasgow and West of Scotland Technical College. In other words, he would venture to submit that the several sections of this Institution would not be satisfied by a single representative on either of these two bodies; inasmuch as he apprehended they had not yet seen the man, even amongst the many able men in their midst, who would be vain enough to assume to a sufficient knowledge of the theory and

practice involved in the five distinct professions that he had named,
to permit of his opinions having much weight at the Council boards
of those Institutions.   Mr Dyer had favoured them with the dictionary meaning of the term "engineer," as follows :—"A constructor of engines; a mechanist ; one who manages a steam engine ;
a person skilled in the principles and practice of engineering, either
civil or military."   Most of them would be prepared to accept this
exposition of the name "engineer" as being fairly accurate ; but
Mr Dyer dismissed the definition as being vague, and asserted that
"from an etymological point of view the name 'engineer' was
wide enough to *include every profession*, and that engineering may be
applied to subjects not usually included in the work of engineers."
Surely this was, he hoped, unintentionally, pretentious, and if admitted led *a priori* to the doctrine that engineers were alike capable
of practising surgery, medicine, jurisprudence, and the like.   Mr
Dyer next proceeded to enumerate no less than what he termed ten
various branches of "engineering;" amongst them he (Mr Williamson)
noticed that he included " naval architecture," "architecture," and
"agriculture."   But as neither of those sciences fell within the
dictionary meaning of "engineer," they could only have been
absorbed, by Mr Dyer asserting, as he had done, that "engineering" included "every profession."   But naval architects, architects,
and others would not, he was sure, agree to be thus absorbed in the
single name "engineer."   They had yet another definition of the
name, for Mr Dyer said on page 131, "For our present purposes, I
will define an engineer as one who has been trained in the theory
and practice of one or more departments of constructive or manufacturing industry, and who is capable of designing and superintend
ing large works of construction or of directing large manufacturing
establishments."   To this also, he for one respectfully dissented as
being inexact, and hoped that other members of the several professions would not consent to be thus eclipsed by the single name
"engineer."   He was quite aware that Mr Dyer's line of argument
was quite within the lines of modern, and, he believed, pernicious
custom that had grown up in public practice.   He meant the custom

of employing engineers to do naval architects' work. He was persuaded, however, that it was not in the best interests of the science of naval architecture that it should be practised by men who had not had the most elementary scientific or practical training in that profession. In that opinion men in the highest walks of the profession had expressed their concurrence. It, therefore, behoved naval architects to speak out boldly against this growing doctrine, and against the continuance of this modern custom which had already led to such disastrous results. He was at one with Mr Dyer in insisting upon, that their most successful men of the future, as in the past, would be those who combined the highest theoretical and practical training. In connection with this, he must express his astonishment at the wail of complaint that went forth from one speaker, and supported by another at the previous night's discussion, because youths were expected to work earnestly at the practical part of their profession, no matter how high their scientific or theoretical education might have been. He had heard those remarks with feelings of regret, as being calculated to encourage youths to adopt a course that would not be for their benefit in after life, and which, if systematically pursued, would make our professional men a body of idealists! He desired to emphasize this the more, because there could be no doubt that those who, like himself, had had experience in this matter could testify that there was what he might call a very decided "kid glove" tendency amongst many of their young men who were training for the engineering and shipbuilding professions. And, therefore, any remarks that would seem to sanction or encourage the growth or continuance of that tendency should not, in his opinion, go out unchallenged from this Institution. In reference to the statement of another gentleman, that masters were unwilling to allow their apprentices to avail themselves of the University and Technical Classes, he might say that his firm (Messrs Barclay, Curle & Co.) had always granted permission to their apprentices to attend such classes; and, being anxious to encourage their apprentices to acquire a theoretical knowledge of their profession, they had made it a rule to pay them for the time spent at

24

those classes, and to allow it to count as a part of their agreed
period of apprenticeship. And in order to encourage those who
wished to attend the evening technical classes, his firm had paid the
fees for them until this year, when it was discontinued owing to the
apprentices not attending regularly and attentively to their studies.
He had reason to know that there were other firms who accorded
similar privileges to their apprentices. With Mr Dyer's views on
the necessity for theoretical and practical training; against cramming for examinations; and on technical education, he most cordially
concurred. And his demand that the newer professions should have
the same privileges as the older ones, was, in his opinion, most
proper and reasonable. Those who had given the subject careful
thought, would also be disposed to agree with Mr Dyer, that the
views of the Edinburgh professor of languages, whom he quoted,
were preferable, viz.: That the students should be allowed "to
specialise their studies according to the bent of their own inclinations rather than be compelled to study all branches of the engineering profession—(he used the name 'engineering' in the broadest
meaning claimed for it by Mr Dyer)- -for it is obviously better to
know one subject thoroughly than to have only a smattering of half-
a-dozen." These remarks were written for delivery at last meeting,
but the subject was not reached that night. Since then Principal
Caird delivered an address at the University, which, in his opinion,
deserved to be recorded in letters of gold. Referring to this subject
he remarked, "A great deal has been said, and justly said, as to the
need for a wider option for studies, in our arts curriculum. I have
frequently spoken on this subject, and I am not going to take you
to-day over such thoroughly beaten ground, but in the few moments
I am to address you, I may point out one or two obvious considerations arising out of the very nature and essence of University education, which should be kept in mind in the attempt to introduce
greater latitude of choice into the course of study. I am happy to
find that since writing the few remarks I am to make that views of
a similar kind have been set forth with much more cogency and
impressiveness in the admirable address of Principal Campbell

Fraser to the Edinburgh Graduates. What occurs to most people on this subject is, that owing to the advancement of knowledge and the ever increasing specialisation of its departments, selection on the part of the student is inevitable, and that as we cannot attempt to know everything, we should be guided in what we select or omit by our individual aptitudes, and by that which those aptitudes govern, the special calling or career in life to which we are destined." He trusted that the effect of the unfavourable comparisons made in the paper under consideration, between the systems prevailing in France, Germany, and America ; also between several of the most important manufacturing cities and towns in Great Britain, and those prevailing in Glasgow, would be to stimulate the merchant princes of the Second City to such action as would place within the grasp of the youth of Glasgow, facilities equal to, if not greater, than those enjoyed by the youths of Dundee and Sheffield, for example, for obtaining a sounder system of technical and theoretical training, both in the University and at the Glasgow and West of Scotland College than at present prevailed. He had shown that in the first portion of his paper, page 130, Mr Dyer enumerated no less than ten different professions, which he claims "as fairly representing the various branches of engineering," including "naval architecture," "architecture," "agriculture," &c., by which he ignored the fact that those professions were practised by the ancients of pre-historic ages, and as far as they knew before even the first elements of "engineering "—using the name in its accepted modern language— were known and practised. They would, however, observe that Mr Dyer abandoned this untenable position when he told them in the second paragraph, page 139, of his paper, "that engineering is not one and indivisible, and that this being recognised it ought to be followed to its logical conclusions;" and he concluded, "that a system which grants a diploma in civil engineering and mechanics, when not even the elements of mechanical engineering and mechanics are taught, and which allows a department of naval architecture and marine engineering to be constituted without making arrangements for imparting instruction in the latter subject, cannot be considered logical." That is precisely my argument; but the above remarks are entirely at variance with the tendency of the first portion of his paper. They were not, however, in a position to judge of these arrangements, as they had not the facts before them, whether a representative of this Institution, however eminent, could have influenced the University Court so far as to have caused them to "disallow" the wishes of the noble-minded and generous donor, who sought to perpetuate the memory of that great man, her late husband, by founding a University Chair for the two professions which he had practised together, with such conspicuous ability and untiring zeal, and which led to such beneficial and economical results to the whole community. It would be ungenerous on their part to demand strictly logical tenets in such cases as these; rather let them be grateful and thankful for the noble qualities of mind and heart that prompted such generous gifts, and hope that in due time and in some way not at present seen by them, other means might be forthcoming that would permit of their strictly logical ideas and conclusions being fulfilled.

Mr W. RENNY WATSON said he was very much interested in the allusions that Mr Dyer had made in his paper to the Massachusetts Institute of Technology. He had for the last ten or twelve years taken a very keen interest in that Institution. He was so impressed with the fact that they had nothing like it in this quarter that he had been advocating strongly the erection of a somewhat similar institution in Glasgow; but knowing that this new Glasgow and West of Scotland Technical College was being formed by the amalgamation of the existing institutions, he had done nothing more in the matter, but when Mr Dyer read his paper, and gave particulars of the work that was being done in that College in America, it occurred to him that he might show them some samples of the work done in it by the lads taught in that College or Institution, and as specimens of the work which the young engineers of America were trained to do. Being of strong opinion that no mechanical engineer could be thoroughly equipped without a good deal of theoretical knowledge, he had tried in more than one case to let lads work in the shop during summer and attend the University Classes in the winter; but he believed that the superior plan was what they had adopted in the Massachusetts College, which was not the only institution of the kind in America. He knew of the Polytechnic Schools in this country, but they gave almost wholly theoretical education, and nothing, for instance, that would assist in the train-ing of the eye and the hand together—a department which was so much attended to in America. So greatly was he impressed by the superiority of the training in the Boston College that he prevailed on his partner to send his son thither for three years. When the lad came home after the first year's work there, he (Mr Watson) was much surprised at the specimens of work he brought with him. He was taught in the knowledge of theoretical engineering at that College, and so many hours were devoted each day also to shop work, the work in the shop was taught him very differently from here—he was instructed by a special teacher, just as if he had been learning to sing, or play, or do anything of that kind. He was not simply put among workpeople, and left to the tender mercies of some good-natured journeyman, but he had an instructor. Now, when he was going through the College himself he saw a lad making six feet of a three-eighth chain. Any mechanic knew that that was not an easy task. These boys were shown how to do every indi-vidual part of the job. They were then left with the iron and everything else necessary in their hands, and each had his anvil and fire to manage, and left to do the work. They had been previously experimenting in hammering on bars of lead instead of iron; and then the heat was the next element in question; and he could assure the meeting, that the boys made those pieces of chains most marvellously well. He particularly observed the perfection of the links and the clean welding. Each lesson in that College was given with the same intention—that whatever the boy was taught to do, the one thing to be aimed at was the nearest possible point to perfection. Everything was taught as an art, not as a thing that might be good enough to pass muster: the lad was expected to do it to perfection. The samples which he had on the table could be seen at the close of the meeting, and they would show what a lad had been able to do after eight or nine months training, of about twelve hours in the workshop per week. It was marvellous how they trained these lads in the schools to appreciate the points in work. For example, he got this history from a lad with regard to a piece of flat filing. He had been aiming at getting 25 extra marks for equal angles, but finding that he could not bring up his work in the time to equal angles, he said he went for the flat filing, and got better marks in consequence, as the lesson was for flat filing. The American was very good in distributing his energies to the best advantage. He thought they would all see that the samples shown were very wonderful for boys of that age. Mining engineers were taught chemistry and mining; and there he had seen lads, each with a laboratory to himself, testing ore. There were five or six tons of ore put down for him to experiment upon for a week, and the whole place put at his command for that time. The boy had first to write out a diagnosis of how he should treat the ore, and then he was allowed to proceed as he pleased, and the result he obtained was compared with the problem he had set himself to do. Next week other five tons were brought in for another lad, and he was allowed to proceed in the same manner, having the benefit of the previous week's experience. At first the College authorities had to buy the ore for the experiments, but now, from the value of the published results gained in the laboratory, they got any quantity of ore from the far West. After lads were under such training for three or four years, they became of great value to employers, and he remembered an instance of a railway magnate, turning to General Walker, of that famous Massachusetts College, and saying—"General, I'll find a place for every one of your present year finished students." The General replied—"I am afraid you won't, as everyone of this year's students are already bespoken." This showed how these boys were available for the management of works, &c. Another instance of how those students were valued came to his knowledge on one of his first visits to that College. A boy was observed sitting quietly as he passed, and whose history he afterwards got.

The boy was there for his second session, and was a capital student. At the end of his first session he had gone home, and applied to the Manager of Colts' Revolver Manufactory for a job. The gentleman inquired where he had come from, and when told the Massachusetts College, he was informed that the gentleman had no use for such students. The lad pled hard and bravely to get something to do, and prevailed. He was set to make locks for muskets or rifles, and in the course of six weeks showed that he was the best workman in the room. After two months' labour in the shop, he was unanimously chosen to examine the work done—all showing how his eye and his hand had been educated together at school. At the end of the year he went to the Manager and said he wanted to go back to Boston to school. The Manager told him he was getting on very well there, and wished him to remain; but he (Mr Watson) saw him beginning his second session at the Massachusetts College. There was no such school in this country ; indeed, except one in Russia, there was not one in Europe ; but he hoped yet to see such an institution in Glasgow. As Professor Woodward, of St. Louis, happily said—" Most schools teach boys mechanics by making something they can sell on the market, such as small machines. Our design is to put the boys themselves on the market ; that is, to make them fit for work." He was much gratified by Mr Dyer's appreciation of the work of these American Colleges, but he could not join with the remark in his paper deprecating the making a fair copy of such a college here as now existed in Boston. He (Mr Watson) did not feel averse to making a copy of a good thing that had succeeded ; and he did not know of a school in this country that had had anything like the success of that of the Massachussetts Technological College ; and he would be very glad if such a school could be started here to-day, without looking for anything better at present.

Mr DYER asked Mr Watson how far he thought the practical training given at Boston could replace the ordinary apprenticeship which was customary in this country.

Mr WATSON said he would have no fear of replacing the present apprenticeship time with such instruction as was given at Boston, except that in that school they, only dealt with material, but not with men. Now, it was necessary to get that other knowledge, for when one began to deal with men, especially Scotchmen—many of whom liked to speak back and have a good deal of their own way—they found the need for that kind of training which no technical school could give.

Mr ALEX. C. KIRK said he would not discuss what was implied in the terms engineer and naval architect, but would leave that to any one who chose to take it up. Nor would he go through Mr Dyer's paper, but would content himself with making one or two very elementary remarks. The education of an engineer began like the education of everybody else, before even the person thought whether or not he was to be an engineer. Now, instead of pointing out how the things should be done, he would point out a few initial defects in education which he had observed in his own experience. He had an examination once or twice a year of his apprentices, in order to choose whom he should take into the drawing office, and he found among them usually several defects. One of these was a defective knowledge of common arithmetic, and that whether the apprentice came from a good college, a famed school, or a Board School. He found that very few of them could do the four rules of arithmetic quickly and accurately ; and, of course, it followed that without the ability to do such sums, as a foundation of education mathematics was nowhere. Then he had observed further that many who came as apprentices, and who should have been taught drawing, knew little about it. Under the present system of sub-division of labour, when girls do the tracing in the drawing office, the boys had not an opportunity of being broken into the use of the drawing instruments, and they came there not being able to draw a clean, straight line. He thought that was a great want in their drawing classes in schools, and he had mentioned it to several, but the South Kensington authorities gave no encouragement to it, and the master had to teach so as to make a living. He had pointed out that the students they sent out could neither make a clean line nor put in the slightest touch of a shade, in order to make a thing intelligible. In that way, when apprentices came to the drawing office they were often of little use for a while. That was generally the case. He also found a good many coming from technical institutions able to calculate, for instance, the strength of a beam, or the momentum of an engine, by rule, but who had very little intelligent idea of what the rule was based on. That was a great defect. He would rather have a young man well grounded in theoretical mechanics, though with little experience of their practical application, than one who had got the rules at his fingers' ends, but did not know whence they came or whither they led. What Mr Watson had shown was done in America was very interesting, and he was sure that the training those lads got in the Massachusetts College must be of great value, more especially as it did not seem to take up very much time. However, it did not take the place of the general training necessary for the handling of men, which was a material item in this question.

Mr W. RENNY WATSON said he fully agreed with Mr Kirk as to the importance of the common elementary studies; and he would like to tell them that the Massachusetts College had a similar opinion, and sought to promote the lads' education by means of a weekly common arithmetic class. The way they managed it was this :—The boys were all in a large hall, the walls of which were covered all round with a black board, and the teacher was in the centre. The boys were ranged in three classes --A, B, C—alternately round the hall. The teacher gave out a count in arithmetic, first to the A scholars, then another to the B's, and the C's got a third count. According to this arrangement, the boys could not copy from each other, as the B's and C's were between the A's, and so on, all round the hall. This showed how carefully the authorities looked to all those little but important details in such a place.

Mr CHARLES C. LINDSAY, C.E., said that there could be but one opinion as to the ability with which Mr Dyer had prepared this paper, and it would be conceded at once that it would be difficult to produce phraseology to which the same meaning would be attached by everybody. Turning to the educational features of the paper, on page 130, Mr Dyer gave a curriculum for engineering students which he had adopted at Tokio College. That was an excellent curriculum, and when coupled with practical experience in summer months, and for two years after completing the course, would, no doubt, be an excellent system for Japan. Mr Dyer proposed the same course for this country. It might answer in the case of some students with means to afford it, but he (Mr Lindsay) did not think it was a solution of the general question of technical education in this country. The Cooper's Hill College authorities had been trying the system for some years, and paid a fee of about £50 for each summer per student, but the work done by the students had not been looked upon seriously by those under whom they had been engaged, but rather as a summer holiday. Doubtless some of the students would derive benefit by this experience, but his impression was that the system had been tried in this country and found wanting. What was wanted was a system which would meet the requirements of the class from which the mass of engineering students and apprentices were now recruited. He (Mr Lindsay) thought that a solution of the question would be found in a combination or affiliation of well equipped technical schools and colleges with the Universities, for the purpose of imparting primary, secondary, and the highest technical instruction, and granting diplomas and degrees. In Glasgow the end might be attained by the affiliation of the Glasgow and West of Scotland Technical College, when thoroughly equipped, with the University for these purposes. The primary technical education could be obtained at Allan Glen's school, before pupilage or apprenticeship, and the secondary education and diplomas at the evening classes of the College of Science and Arts and Anderson's College during apprenticeship, and the highest technical education and the degrees at the University, the attendance of eminent students being facilitated by bursaries and scholarships. That seemed to him the best solution of the general question. With regard to the suggestion on page 156, that an experimental tank should be instituted at the

University, he could not give a decided opinion upon it—it was a question for shipbuilders. Would shipbuilders on the Clyde submit the particulars of the vessels on order to any one man for experiments ? He thought not, but if they would, the Professor would be exclusively engaged on these questions. To students an experimental tank without questions leading to practical results would be utterly uninteresting. The tank at the Messrs Denny's work cost about £5,000, and it took about £1,000 a year to keep it up. Were no income derived from it, that meant a capital sum of £25,000 to £30,000. That along with the doubt as to the business patronage of the shipbuilders, seemed to him to be against the proposition. He believed that such tanks should be left to private enterprise.

Mr IMRIE BELL would like to say a few words on what Mr Williamson had said with regard to the status of the naval architect. He was afraid that gentleman was a little narrow in his remarks when he averred that the status of a naval architect might be lowered by being called an engineer. Now, the Institute of Civil Engineers of London were very particular as to whom they admitted, but they all knew that engineering embraced almost everything connected with the Arts and Sciences; and his late brother, Mr R. Bruce Bell—one of the former Presidents of this Institution —was very anxious that the shipbuilders should be allowed to be incorporated with the members of that Institution, and he succeeded greatly, and now many eminent shipbuilders were members of the Institute of Civil Engineers of London. He did not understand how any one could think that a naval architect would be lowered in status in being called an engineer. He would like to see them altogether designated as engineers under the various different denominations. The Institution, as founded under Telford, was established for the general advancement of mechanical science, as well as for civil engineering. He thought that it would be of great advantage if all the various branches of engineering could agree to federation, as unity is strength; and though each branch might act independently as to distribution of funds and social organisation, they might all centre under one head. There were many serious difficulties in the way which would have to be calmly considered, but he merely suggested that it might be most beneficial to have all the various shoots of engineering issuing from a main trunk—engineers. He would state them alphabetically—namely, civil, electrical, gas, hygiene, marine, mechanical, mining, and water; all combined in one Institution, where they might have a common public hall or theatre, and reading-room with separate libraries. If this could be carried out, it might be at less expense to the individual societies, and to the advantage of all.

Mr WILLIAMSON said Mr Bell had misunderstood his remarks. He didn't consider it derogatory to be called an engineer, but he wished to say that naval architects did not want to be swamped by the engineers.

Mr W. RENNY WATSON said that a winter or two ago he had met a very famous engineer—James Naysmith—who told him (Mr Watson) how that once when in the witness box in London, a question had turned up, regarding which one of the Counsel said—"O no, we cannot go into that—it is a question of engineering." The judge, however, questioned Mr Naysmith as to what an engineer is, and the reply he got was—"I submit that an engineer is a man who applies common-sense to the proper use of materials." He (Mr Watson) thought they would find it difficult to get a better definition of the term.

The CHAIRMAN intimated that several MSS. had been received bearing on the subject, which are subjoined.

Remarks received from Mr G. C. THOMSON, President of the Graduate Section of the Institution :—

At a special meeting of the Graduate Section of this Institution, held on the 19th April, called for the purpose of discussing the subject of the Education of Engineers, as brought before the Institution by Mr Dyer, the following remarks were laid before the members, in order to open the discussion :—

In November, 1883, Mr R. L. Weighton, in his Presidential

Address to the Graduate Section, entitled "The Prospects of Engineering and the Engineer of the Future," stated his belief that there was a period of activity and progress in the future for the engineer, whether civil or mechanical, that his labours were divided into two spheres—viz., dealing with matter and using methods; also that progress would be in two directions—by improvement in the qualities of materials, and by increased intelligence in the applications of methods. He enumerated among the methods a better acquaintance with the principles embodied in those branches of knowledge applicable to the work of the engineer—viz., mathematics, pure and applied, geometry, natural philosophy, chemistry, and geology.

In addition to this theoretical knowledge, there was necessary a practical acquaintance with engineering details, and indications were not wanting to show that this indispensable practical knowledge was in the near future likely to be combined with first principles. Also that the engineer of the future, while worthily keeping up what had been done in the past, would strike out into new and successful paths for himself.

In a paper read two years later, I stated that while agreeing with this view of the question, I thought that still more was required to become a successful engineer. For, let a man be ever so well up in the theory and knowledge of the practical details of machinery and works of engineering; yet, if he has no power of organisation, and no tact in dealing with his fellow-men, whether in the character of customers or in that of employés, he will inevitably fail to become successful, despite his knowledge, earnestness, and perseverance. Some may not be inclined to admit this, but a little thought will show it to be the case, and observation will only prove it more forcibly.

The foregoing gives the gist of all that has been said or done in the Graduate Section on the question of education.

The keenness of foreign competition in many, if not all, branches of manufacturing industry, has awakened a strong desire to know the reason of the loss of trade with other countries, and, if possible, to regain our former supremacy as the workshop of the world. On the Continent of Europe and in America, our chief competitors, nearly every State has its technical school or college, in which the education given is of such a nature as is supposed to fit the scholar for the particular business he intends to follow. In this country, and more especially in our own city and district, this idea is taking tangible shape in the formation of the Glasgow and West of Scotland Technical College, and the Council of this Institution, as you are aware, has sent a representative to the governing body.

Mr Dyer has gone into the question very fully, and his paper is a valuable one, and I shall only make a few remarks on some of the points suggested by the reading of it. On page 129 I quite agree with his definition of the word engineer, and taking civil engineering as including all that is not embraced by military engineering, and as it is impossible for any one man to know all this and practise it, his classification, as given on page 130, is subdivided into very convenient branches, each of which is quite enough for one man as a speciality, and the preparatory training for any one of these is common to all.

On page 131, however, where Mr Dyer defines an engineer as one who shall be capable of designing and superintending large works of construction, or directing large manufacturing establishments, and leaves out the foremen and higher class of workmen, I am not at one with him. The work done by Bessemer, Siemens, Sir William Thomson, and Young, was the result of the combination of great practical and theoretical knowledge, and was only successful by reason of their great practical skill. On page 132 I quite agree with Mr Dyer in regard to examinations as usually conducted, as it is usually a case of cramming without understanding, and leaves the scholar worse than before, and with all the spring taken out of him, if carried to excess.

On pages 136 and 137 I must say that I side with the engineering professor, and not with the theological professor, but at the same time I do not think the present college course is the best possible. Pages 139 and further on, I do not think that a special college with workshops attached, such as are sketched, are necessary, as by the appointment of a few more professors to lecture on special subjects, the needs of each case could be met.

In regard to the training as sketched on page 152, I shall refer later on. On page 153, I quite agree that it is an essential that the health of the pupil be cared for as much as the mind and skill, as without it there is no chance of succeeding in any business in these days of keen competition. On page 157 I do not see how the practical training as there sketched could be well carried out, even though it would suit the Scotch Universities. Page 158, training of foremen, &c., is good, but leaves one with the impression that too many difficulties might be thrown in the way. In regard to languages, they are not an absolute necessity in order to be a good engineer, but they are of great commercial utility. I am in favour of having special schools for some particular crafts, such as are indicated in page 159. I quite agree with Mr Dyer's remarks from page 160 to the close, and trust that as time advances the picture will be brighter.

I shall now proceed to give a few results deduced from observation while employed in various capacities. At one time, while in charge of a drawing office, I had three journeymen trained in the polytechnic schools of the Continent, and who had had some experience in the shops and offices. An M.A., who, after taking his degree, had served his apprenticeship as an engineer, and had been a short time a journeyman, and an apprentice about two years at his trade, and who had had only what is commonly known as the three R's in the way of schooling. All of these, save the M.A., whom I have lost track of, at the present time occupy good positions of trust and responsibility, but I found that despite their extra training and knowledge, they were not of much value to me for getting out the work, and they were very slow, in fact the apprentice did best of them all. Again, in the shops where I have wrought alongside of foreigners who had received their training in technical schools, they were unable to compete with their neighbours save in rare instances.

Among chemists again, in the German Universities, I have observed that those students who had come from the polytechnic schools in order to make an investigation and take a degree, were nowhere compared with those students who came direct from the school to the University. After a short time the students could beat the technicals hollow for manipulative skill, accuracy, and observation, and in having a good idea of going about their work. In the same way I observed that of the foreigners, such as British, Americans, &c., those who had learned a mechanical trade or profession before beginning the study of chemistry and allied subjects, possessed a great advantage over those who had not had this training, and who had begun the study of a science requiring keen powers of observation and great manipulative skill after they had attained the years of manhood. It seemed almost an insuperable difficulty to acquire the delicate touch and power necessary. The want of self-reliance was much more noticeable among the technicals than among the students. Of course there are exceptions to every rule, and I have known exceptions to what I have stated above.

I am of opinion that whether the student or apprentice will be a manager or a workman only in after life, his preliminary training should be practically identical, as it is not given to us to know beforehand what we shall eventually become, and there is always the necessity laid upon us of being able to earn a livelihood, at the close of our apprenticeship, by the exercise of the craft which we have learned, and it is only in rare cases that this does not hold good. It is an immense advantage to the master or manager of a large factory if he can set to at any time and show any one of his employés how to do the work required of him, and to know when going through the works if the work is being carried out properly, in a workmanlike and economical manner; and unless he is a good workman himself he cannot do so.

Before beginning with the education of the apprentice in the works, I would like to say that each boy and girl should have a fair, good, sound education. The object not to be, as at present seems the case, to cram as much as possible into the pupils' head, in order that they may pass given standards at given ages, but to teach them to use their faculties of reasoning, observation, and thought, so as to develop self-reliance and to use their judgment. At the same time the parents or guardians of the pupils ought to observe the indications given by the pupils, not so much by what they say but what they do, in order to discover the bent of their minds, so that they may determine what path of life will suit them best.

This view of the question, however, seldom receives the attention its importance deserves, and we see the consequences daily in the failures of many lives which otherwise were full of promise and brightness. I am sure you all know instances of this.

Allowing that at the age of 16 the scholar has had a good grounding in mathematics and natural philosophy, and is ready to begin his apprenticeship. As you will see, I am not in favour of doing away with the apprenticeship system, I would make it stricter than at present, more approaching to what it was in the time of the guilds, when the apprentice learned all his master could teach him, and then qualified himself by travelling and working in other workshops to be a master himself if he wished to do so and could comply with the conditions laid down, and show his skill as a craftsman. This would require modification in these days, and it would meet the same end if the apprentice were allowed away an hour or two sooner to attend certain classes, attendance on these classes to be compulsory, and from time to time examinations made to test his knowledge, and finally before he receives his indentures he must pass an examination that will show his skill as a craftsman and his theoretical knowledge are up to a certain standard such as will enable him to earn a livelihood and to go on learning more and more of his profession. The average period of apprenticeship I would still leave at five years, as that is short enough to turn out a good tradesman. The examinations should not be fixed for a certain period of time, and time and material should be given to him so that the conditions would be as near as possible what he would meet at any time in the course of his career.

Situated as we are in Glasgow amongst workshops of all kinds, there is not the slightest need of a special workshop being erected

26

for the apprentices to learn the use of tools, as the actual workshops are infinitely superior for this purpose, while the lectures and classes he would be attending would treat of the reasons for what he would see about him every day in the works.

During the last year of his apprenticeship it would be advisable to allow the apprentice either one day per week, or the last few weeks of the year, to attend a special laboratory, where he could try the testing of materials and machinery under the direction of the professor, this laboratory not to be employed as a school for learning the use of the ordinary tools, but solely for the purposes of testing material and machinery of various kinds. He would in this way be fitted to derive the utmost benefit from the practice after seeing and working among the machinery in the shops, and having acquired the necessary hand skill. If he chose he could then attend the purely scientific studies at the University, and his workshop experience would be of untold benefit to him ; or he could remain as a workman and be all the better of his scientific and necessary training. In this way the bodily strength of the pupil is attended to without interfering with the development of the theoretical department.

By a little management on the part of the employers and their foremen, and that of the professors or teachers, this could be arranged without interfering with the work being carried on in the shops, and by the time the pupils were 21 or 22 the course necessary would be finished. Those of course who required to begin work at 14 years of age, on leaving school, would not have such an advanced knowledge, and would have to be content with a lesser amount of theoretical knowledge, but the same amount of hand-skill, before receiving their indentures at the end of five years. In their case also, the evening classes would require to be attended, but it would require attendance at classes while journeymen to enable them to attain the knowledge necessary to take the degree of Engineering, if such were instituted.

This refers to the testing laboratory previously mentioned. One idea that might be carried out for such of the apprentices as wished it, would be that of making each one take his turn as foreman for the day, set the work for the others, and see to its being carried out. This I would keep as a prize, so to speak, in order to put the lads on their mettle, and from their behaviour in this class a fair indication of their organising power and tact in dealing with others would be gained, the professor or some other specially appointed man of good practical experience being on the ground and overlooking the carrying out of this idea.

This would be a good opportunity for those successful engineers who had made their fortune, of employing a portion of their well-earned leisure in furthering the science which had been their life-long companion, and I venture to say that it would be a pleasant one, and at the same time much sought after.

The necessity of an engineer knowing at least one foreign language is evident from a commercial point of view, and master of one language, it is easy to acquire another.

Mr Weir's motion is practically a part of the paper, and cannot rightly be discussed separately, only it is but reasonable that a representative from the Institution should be sent to the University Court in order to temper the deliberations of the learned professors with a modicum of practical knowledge, so that any scheme put forward would not interfere with the executive skill and management of the manufacturing industries.

Although I have read the paper over two or three times, I would not like to say that I have fully understood it in all its bearings, but as it affects the Graduates more than any others, I trust you will give your opinions on the matter freely.

In the course of a very full and animated discussion the following points were prominently brought forward; generally the views I have expressed were cordially agreed to.

That the apprenticeship system ought to be maintained, but that the apprentices should not be confined to one branch or machine during the whole time, and that they be allowed away a little earlier to attend classes; attendance on which should be compulsory, and which would form a part of the training of the pupil. The system of studying in winter and working in summer would not work in this country.

Opinion divided between the advisability of going first through the shops and then the college, or *vice versa.* Strong opinion in favour of extramural teaching, and that the universities be thrown open, so that any one can demand to be examined for a degree, no matter how or where he may have acquired the requisite knowledge to enable him to do so. Some even went further and thought the Government should appoint examiners (such as Board of Trade), so that no matter in which town a candidate found himself, he could get his degree if capable and so minded, and also that there should be varieties of degrees, for instance, as a mechanic, as a manager, and for scientific investigation, which would give the holder a certain status, as universities were apt to be stereotyped and not open to progress.

Examinations as at present conducted were generally condemned on account of their not being a real test of the candidates' knowledge, while a few thought that they gave a stimulus on account of the pecuniary gain, and thus gave rise to a love of learning for its own sake, which otherwise would have lain dormant.

In any case first have a good grounding at school as long as parents can afford, then into shops and evening classes, and then, if he can afford it, to the University.

Some who had been in the technical workshops condemned it as practically wasted time, and all were of opinion that the real workshops were the best.

The general opinion was that the paper seemed to incline to make engineers without passing through the shops, and to make the way more difficult for those who had not gone through a technical course of attending the University and taking a degree.

Further, that as in the past, most of the masters at present had been workmen who had risen, it would not be advisable to throw more obstacles in the way of others rising for the sake of training the few who could rest assured of becoming masters shortly after completing their training in the technical schools. Further, owing to the fact that in actual work every day the questions to be answered, and the work to be got through, did not admit of the slow, deliberate methods of the schools where time was of little account, but must be done in a few minutes, hours, or days, as the case might be, and under varying conditions, that the training for every engineer should be practically the same in its preliminary stages but carried further as the ability and inclination of each would allow.

Text books in their proper place are good, but used as they often are, are productive of great and lasting harm to the student.

I think this covers all that was said at the meeting of the Section.

Mr GEORGE C. THOMSON said that with the motion proposed by Mr Weir, and so ably seconded by Mr Hyslop, he quite agreed, especially in regard to the thoroughness and practical nature of the examinations proposed, as at present they took no cognisance of the skill as a craftsman, which was a point of very great importance. One of the speakers had referred to the drudgery of serving an apprenticeship to the practical work, but where the thirst for knowledge was properly cultivated, it would be found a pleasant, interesting, and profitable study to observe the deductions of theory carried out in practice, and to note the corrections requiring to be made before they wrought in unison. It would be a poor day for engineering if the idea that serving an apprenticeship was unnecessary drudgery got the upper hand. He was certain that if employers would only pay more attention and be stricter with their apprentices during the time they were with them, and see that they were really fitted for the work they had to face, they would be more than repaid by the better quality and increased quantity of the work turned out, and it would tend greatly to the increase of more cordial relations between employers and employed. One point that had not been touched upon in the paper under discussion was that of paying attention to the appearance of the work or machinery turned out, because as it was a matter of money-making, the designer who could turn out a beautiful machine would sell far more than the maker of an ugly one, although for practical use they were equal. In fact, the one commanded the market, while the other had to go begging. This had had a good deal to do with the success of foreigners when competing against us, and ought to be attended to far more than was done at present. Perhaps this was what Mr Dyer referred to when speaking of a liberal education in art; if so, it was as essential for the workman as for the master to be highly trained in this respect. If the education of engineers was carried out on these or similar lines, we could defy foreign competition to wrest from us any trade worth having, and still retain our supremacy as the workshop of the world, aided as we were by abundance of raw material, good and cheap labour, and a climate which could not be surpassed for the endurance of intense sustained labour without harm to the labourer. While listening to Mr Ward reading his Memoir of the late Mr Denny, the thought had struck him that it was the best answer to those who sought to do away with the apprenticeship system, and substitute a course of technical training such as is carried out in Germany, France, and America. The liberal and sound education enjoyed by the late Mr Denny before going into the workshops to learn his handicraft, enabled him to see to what particular branch he required to devote most attention, and as a master over the details of his business, with the result that his name had become a "household word" wherever ships were known, or the interests of capital and labour were concerned. His life and works pointed out the lines on which we should go, and if we learnt the lesson taught thereby, we should not need to fear the competition of foreigners, or even of our own colonies. It was good to have a high standard of excellence, even though we fell short in the endeavour to reach it; and in this way Mr Dyer's paper would be of great and lasting benefit to the cause of education, on account of the interest awakened thereby on this subject.

Mr T. ALEXANDER, C.E., writes:—

I wish to give a short sketch of the working of the department of Civil Engineering in the Imperial College of Engineering, Japan, during the seven years that I was professor in that department. Students entered the department after two years' general study in the College, suited to the requirements of all the students. During the first summer session they were taught Surveying, and had a short lecture course on Applied Mechanics suited more particularly to the requirements of students in the other Technological Departments; to them only introductory to the more complete winter course. In like manner the C.E. students attended suitable portions of the courses of lectures on Steam, Mechanism, and Mining by the professors in the departments of Mechanical and Mining Engineering and of Naval Architecture. In the winter session they had four hours per week lecture work on Applied Mechanics; and what I particularly wish to call attention to, six hours office work, real work, designing from the principles taught; beginning with ties, struts, and simple framed structures, and going on to more complex work, giving the most recent graphic representation of the states of strain, under the constant supervision of the professor. They had besides four hours per week office work sketching details from the most recent practical engineering works with which the library was constantly supplied. In the second summer the students projected a line of railway a few miles in length, generally to the order of the Chief Officer of the Government Railway Department; in this way alternative routes of all railways near Tokio were made. In the winter the students plotted their *railway work*, and made the designs for stations, bridges, tunnels, and other works. They had a lecture course on *Civil Engineering* from two professors, who divided the work into the more theoretical and more practical parts; the theoretical principles being applied in the office to designs of earth works, and of stone and iron structures. This completed the two years' technical course, and they now entered upon a two years' practical course. The first summer they spent in the Government Workshops at Akabane, and in recent years in the *Engineering Laboratory and Workshops* within the college; they attended from 5 a.m. to 3 p.m., and were taught the *use of tools* under skilled workmen, made experiments on the strength of materials, took indicator diagrams, and saw in full operation machine tools, dynamos, and the drawing out of ships' lines, model making, &c. For fifteen months the students were now apprenticed out on works—railway, harbour, lighthouse, river improvement—in every part of the empire; reports and sketches being sent to the college. These students had with them copies of Stevenson's and Harcourt's works on rivers, canals, harbours, &c., and read prescribed portions. The last six months were devoted to *diploma work*. The students, being placed in the conditions of actual practice, were required to make complete and finished *designs* of an engineering work, to write an *essay* on the same subject, and pass an examination on theory, where books were freely allowed.

The graduates in civil engineering, who numbered to about seventy in those seven years, got immediate and well-paid employment in the different local boards of works, and in every department of the Government. Two obtained employment upon railway work in America, which they have retained for four years, and one is at present in paid employment on the Forth Bridge works.

This last point is one upon which I have a strong opinion—that students of engineering should be so taught, or rather trained, that the diploma which they receive may readily obtain them employment with remuneration upon which they can live.

Sets of plans, executed by my students, illustrating the course in civil engineering of the Imperial College of Engineering, Japan, were exhibited in the Educational Department of the "World and Cotton Exhibition" in America a few years ago, and obtained honourable mention. Sets were presented to the Madras Engineering College and the Steven Institute, and met with approval. I should be glad to show them to any one interested.

I append my last prospectus in Civil Engineering and Applied Mechanics. (See Appendix B.)

Mr R. F. MUIRHEAD, M.A., B.Sc., writes:—

In response to your kind invitation to me to take part in the discussion on Mr Dyer's paper "On the Education of Engineers," I have much pleasure in sending the following remarks. As one of Glasgow's earliest graduates in engineering science, I may have some title to speak on this subject, though lack of experience as a practical engineer will make it proper for me to deal chiefly with those parts of the paper which have special interest for the educationist as such.

One valuable feature of the paper is the acquaintance it shows with the state of technical education in other countries. From the information supplied by the paper on that head, it is perfectly clear that our own country has still very much to do. The existing conditions of the industries are such that the old system of technical education in the workshop has been practically destroyed, and we have not in this country attained to anything like an effective substitute. Hence the great and even national importance of such schemes as that of the Glasgow and West of Scotland Technical College. It will need the best knowledge and brain power of Glasgow to solve the problem that lies before such an institution. But this aspect of the subject is not now, I think, in danger of being neglected, I shall therefore pass to another aspect, which bears on education generally as well as on the training of engineers.

On page 132 of Mr Dyer's paper there occurs a very pertinent quotation from Professor Bryce, on the dangers of our examination system. All educationists will admit the justice of these remarks of Professor Bryce. Competitive examinations are certainly an evil, though perhaps at present a necessary one. I say perhaps, for it is quite conceivable that the ill effects of examinations, in stimulating cramming and superficial knowledge, in destroying the freedom of the teacher and in repressing the individuality of the student, may more than counterbalance the gain of a stimulus to exertion, and an impartial (though also inaccurate) criterion of merit. But I should like to point out certain modifications which would deprive our examination system of some of its chief defects.

First.—The examination ought to be, as far as possible, a piece of real work, performed under the conditions of real work. Thus, in written examinations, works of reference should be allowed, and the

27

questions set should be of the nature of actual problems which might occur in practice. But a still more preferable test is a piece of work set to the student, or chosen by him, as in the case of candidates for degrees at the German universities. The German candidate must do a piece of original work, which is called the "Arbeit." It may be, *e.g.*, a literary or philosophical thesis, or a mathematical essay, or an experimental research. It is done under the conditions of any other piece of real work: the time is unlimited, all available helps are allowed, only they must be duly acknowledged.

If examination is to dominate teaching, how much better would it be were it dominated by such a test, rather than by a written examination stereotyped for a hundred other candidates.

Under the present system, dominated by the written examination, a subject is got up with the view of retaining as many propositions as possible in the memory, to be reproduced on paper under high pressure. Were the other system followed, as in the German universities, a subject would be got up in order to *use* it. The temptation to cram would be gone. In fact there would be all the difference between getting it up for *show* and getting it up for *use*.

Second.—Even where the examination by questions is retained, it ought to be, whenever possible, adapted specially to the individual student, so as to really find out what he knows, and not simply to test his conformity to some arbitrary standard of information or acquirement. This principle is adopted in Germany also, at the universities and schools as well as in the State professional examinations. These two principles adopted would do much to improve the present enormously wasteful system of examinations.

Many other important points call for remark, but I must be content simply to express my high sense of the value of Principal Dyer's paper in ventilating this important subject.

Mr SINCLAIR COUPER writes :—

I am very glad that an opportunity is now afforded, by means of Mr Dyer's paper, for the discussion of so important a subject as the

Education of Engineers. In my own case, like many others, I completed an apprenticeship of five years, and sometime afterwards attended Glasgow University, taking the course there prescribed for engineering students—an education which in all covered a period of somewhat over seven years. I do not think it possible that a thoroughly practical and a theoretical education can be given in any shorter period. There is a tendency now with some to decry the apprentice system, and to imagine that all the practical education that is required may be attained by short periodic visits to a workshop or office, and passing a few hours each day within its walls.

Should we ever have a system of engineering education laid down on regular lines, and carried on in well-equipped institutions, then a youth who proposes to become an engineer could, after leaving school, take up for a time, say two years, the special subjects connected with the profession he has chosen, studying them in such an institution and on the lines laid down. In this case the regular apprenticeship might be reduced to a period of four years, because in addition to the theoretical and scientific training which the apprentice would have, he would also have partially acquired a knowledge of the use of tools and machines. But on no account should the apprenticeship be done away with; for by no other method can young men see how work is carried on in a regular and scientific way. Nor would it be wise, I think (with all deference to the opinion of Mr Dundas) to attend practical work for only a portion of each day. To do engineering work of any kind thoroughly, and to follow the processes of constructive or of office work through all their details, one must attend regularly from day to day, and from the hour of starting in the morning, whether that be six o'clock or later, until the hour when work stops for the day. In the same way only can one see how bodies of men are managed and controlled. To learn the best methods of dealing with men, one must associate with them for regular and unbroken periods.

Nor do I approve of working in the summer season, and studying in the winter. The engineering course of study should continue all through the year, with short intervals of rest every three or four months, instead of, as at present, five months of work, and seven months of rest. In this way time would be saved to the student. During the period of apprenticeship, there is nothing to prevent the apprentice studying, on some evenings of the week, subjects bearing on the practical work he may be engaged in at different stages of his apprenticeship, and also to prepare for taking the finishing courses of his studies at the University with the view of obtaining a degree or other mark of qualification that might be instituted. By such a course the benefits of the apprentice system might be retained and added to that higher education, which, it is to be hoped, will soon be placed within reach of all who desire it. Some six years ago I was engaged for some time in a large ironwork in the Nord district of France, and had good opportunities of observing how work was carried on. There were two classes of men, the greater consisting of the ordinary workmen, who appeared to have little or no technical education. These worked in gangs; each gang or squad was controlled by a young man (he and his compeers forming the second and smaller class) whose headquarters were in the drawing office. These young men were highly trained, and educated in the scientific and higher branches of engineering and mathematics, but had comparatively little practical knowledge, and certainly in most cases none at all of how men should be managed. Under this system work was undoubtedly got through; but not in a fashion that would have satisfied a Briton. In course of time, the differences between these two classes will certainly diminish, as Mr Dyer states in his paper; but I think it would be a mistake were we to institute any system which might tend in the least degree to produce such differences among ourselves.

Mr DYER, in reply, said his paper had caused a long and interesting discussion, into the details of which it would be impossible to enter; space would only allow a few of the more important points to be noticed. They were much indebted to General Walker, and the members of the staff of the Massachusetts Institute of Technology, for the interesting information which they had given about that institution, which was a model of its kind. Mr Watson misunderstood him when he said he deprecated making a fair copy of such an institution as that at Boston. By referring to the paper it will be found that when he said, in organising engineering education in Glasgow, we should not imitate any existing institution, he did not allude to that at Boston specially, but simply meant to say that whatever is done here should be of the nature of an evolution of existing institutions and arrangements, and not a mere cut and dry importation. It is possible to have this evolution, and still have all the essential features of such an institution as that at Boston. With regard to the practical work done by the students to which Mr Watson alluded, he (Mr Dyer) had not entered into a description of it, because it was altogether distinct from that done in the School of Industrial Science of the Massachusetts Institute. In connection with that establishment, a School of Mechanic Arts has been established, in which special prominence is given to hand work along with high school studies, affording an opportunity to such students as have completed the ordinary grammar school course, to continue the elementary scientific and literary studies, together with mechanical and freehand drawing, while receiving instruction in the use of the typical hand and machine tools for working iron and wood. The general plan of the school is similar to that of the Imperial Technical School of Moscow, the Royal Mechanic Art School of Komotau in Bohemia, the Ecole Municipale d'Apprentis of Paris, or that of the Ambachtsschoole of the principal cities of Holland, but somewhat modified to suit the different conditions existing in America. Such a school is intended in great part to replace, or at least to shorten, the ordinary apprenticeship, such as exists in this country; and evidently Mr Watson is of opinion that to a large extent the training there given is sufficient to replace that obtained during apprenticeship. On some other occasion he might give an account of schools of this type, but in the meantime he would point out that if they were adopted in this country, they might get over almost the only difficulty which has been offered regarding the proposals made in the paper—that is, the employment of the students in the summer time. If we had an organised College such as described in the paper, and a School of Mechanic Arts similar to the one in Boston, the college students could attend the workshops during the summer recess, while those of the school of mechanic arts (who would be content with a much lower standard of theoretical training) could attend some elementary classes during the summer session of the college. In the absence of such a school, the manner in which the students should take their course of study might safely be left to the decision of circumstances. The richer students would no doubt take it in the order mentioned in the paper, while the poorer would first serve their apprenticeship and attend evening classes, the best of them afterwards taking a session or two at the day classes. Mr Williamson's remarks about the meaning of the term "engineer" were sufficiently answered by those of Messrs Bell and Watson. He has, however, further strangely misunderstood the paper when referring to the engineering department of Glasgow University. He (Mr Dyer) had on several occasions spoken warmly in praise of the generous donor alluded to, but his remarks in the paper had not even the slightest allusion to her. What he complained of was that the University authorities should institute a department of Naval Architecture and Marine Engineering, and make no arrangements for teaching the latter subject. That either the name should be changed, or the subject should be taught, is surely a very reasonable remark to make, especially as in Glasgow, if the department were properly organised, it might be made entirely, or very nearly, independent of endowments. Mr Duncan has congratulated the author on having the courage to break the ice covering a set of subjects of more importance to the whole of them than even the merits of steam jackets, or the latest departure in valve gears. He must return the compliment, and congratulate Mr Duncan as the only employer in the Institution who has paid any attention to the most important part of the paper, or at least who has had the courage to express his opinions. If our late President had been spared, he certainly would have had something to say on this part of the subject, as it was largely due to his influence and example that the author's thoughts had been directed to the social questions affecting the capitalist and the labourer. For a considerable time past Mr Denny had been in the habit of sending him reports of his public speeches, and from these, and from correspondence and conversation with him, it was quite evident that the main ambition of his life was to realise ideals which are generally thought to be unattainable—Carlyle's Industrial Captain and Regiment—yet living with a greater amount of liberty, equality, and fraternity than is usual. He is gone, but it is hoped that others will arise who will carry on his work, and that his example will incite many to devote themselves to the solution of the great social problems with which we are confronted. This must be done in no dilletanti manner, but with earnestness and thoroughness, and largely in the spirit of the great Scottish seer named above, whose writings on this subject have for more than forty years been ignored by most of those concerned. Among other many wise things he said:—"The main substance of this immense Problem of Organising Labour, and first of all of Managing the Working Classes, will, it is very clear, have to be solved by those who stand practically in the middle of it; by those who themselves work and preside over work. Of all that can be enacted by any Parliament in regard to it, the germs must lie potentially extant in those two classes, who are to obey such enactment. A Human Chaos *in* which there is no light, you vainly attempt to irradiate by light shed *on* it; order never can arise there." . . . . "I anticipate light *in* the Human Chaos, glimmering, shining more and more; under manifold true signals from without That light shall shine." . . . . . "To be a noble Master among noble Workers will again be the first ambition with some few, to be a rich Master only the second. How the Inventive Genius of England, with the whirr of its bobbins and billy-rollers shoved somewhat into the backgrounds of the brain, will contrive and devise not cheaper produce exclusively, but fairer distribution of the produce at its present cheapness. By degrees we shall again have a society with something of Heroism in it, something of Heaven's blessing on it!"

The CHAIRMAN then stated that the motion proposed by Mr Weir, and seconded by Mr Hyslop, was now before the meeting, the terms of the motion being as follows :—" That in prospect of immediate legislation affecting the Scottish Universities, the Council be authorised to take action with a view to securing the direct representation of this Institution on the governing body of Glasgow University."

The motion having been put to the meeting, it was unanimously adopted.

The CHAIRMAN proposed a hearty vote of thanks to Mr Dyer for his paper, remarking that although it had not called forth quite unanimity of opinion, they could not have expected that it should do so, but all the same it was a most valuable paper to the Institution.

APPENDIX A.—From General WALKER.

CIVIL ENGINEERING.

The course in Civil Engineering is designed to thoroughly train the students in the principles upon which the profession is based, and to qualify them to occupy technical positions of trust and responsibility as soon as they shall have gained the necessary familiarity with the practical details of the particular branch of the subject into which circumstances may throw them. The principal point which is kept in mind in the course is that principles and practice must go together, and that the student must be made to clearly see the relation between the abstract principles which he learns and the actual every-day work of the engineer, without, however, devoting much time to minor details of practice which he will learn much better and more quickly after he leaves the school. The rapid specialisation of the various departments of Civil Engineering has furthermore rendered it necessary, in order to cover all of them adequately, to allow the student a certain degree of freedom in the choice of studies, within certain limits, and under the direction of the Faculty.

In the second year a thorough course is given in surveying, including extended practice in the field, and instruction in the use and adjustment of the various instruments, such as the compass, level, transit, solar compass, &c., The astronomical operations involved in determining the meridian are made familiar, and practice is given in the various branches of land, city, subterranean, hydrographical, and topographical surveying. Together with this work in the field—which is done in small parties, and under the direct superintendence of an instructor—goes the work in the drawing room, which consists in the representation upon paper of the results of the surveys, the study of conventional signs, and of pen and coloured topography. The object in this course is not only to make the student familiar with the operations involved in surveying, but to make him actually able to perform those operations with the necessary accuracy. The school has a large collection of instruments representing the best makers, and a full outfit of minor instruments employed in preliminary surveys and reconnaissances.

In the third year the student approaches the more purely technical work, and is carried through a course on Roads and Railroads, including the survey, location, construction, and maintenance of their route of communication. With the lectures and recitations on the subject goes work in the field and in the drawing room. The former consists in the survey of a line of railroad several miles long, with the necessary calculations of earthwork and of cost. The work in the drawing room consists of the study of paper locations, with practice in locating roads upon suitable contour maps prepared from actual examples. The student is further required to make drawings of one or more structures from measurements taken by himself. The work in surveying is continued, and the use of the plane-table and of the stadia is taught, with extended work in the field. A course of lectures on Foundations, and one on Stereotomy, are also given in the third year, as preparation for the more advanced work in the fourth and last year of the course.

The fourth year is devoted almost entirely to purely professional subjects, including the following:—

Bridges and Roofs.—The calculation and construction of structures of this kind are taught in a very extended course, and the object in view is to enable the student to compute and design the ordinary structures in use. The different methods of computation are carefully gone over, and the practical arrangement of parts, grouping, proportioning, and comparison of details forms an important part of the course. Each student is required to make complete designs and working drawings of one or more structures of iron, with estimates of weight and cost.

Principles of Construction.—In this course the student is carried through the principles governing the computation and construction of all the different engineering structures, such as retaining walls, piers, abutments, buttresses, masonry, dams, arches, centres, foundations, &c. It includes a thorough course in graphical statics, with practical examples in the drawing room.

Hydraulic Engineering and Hydraulics.—The course in theoretical hydraulics includes an extended study of the laws of hydrostatics, and of the flow of water over weirs, in open channels, and in pipes ; also of the methods of hydrometry, which are illustrated by practical works in the field, a large collection of the necessary meters, floats, and tubes, having been provided for the purpose, the students are made practically able to measure the volume of water flowing in any sort of channel, and to work up and discuss their results by the most approved methods.

Under hydraulic engineering, a study is made of the subjects of water supply, water power, hydrology, rivers and canals, harbour and coast works, pumps, &c., combined with the practical designing of hydraulic works, such as a system of water supply for a town, based upon actual cases, the data for which have been collected from various engineers.

Sanitary Engineering.—In this course the plumbing and draining of houses, the treatment and disposal of sewage, the works of sewerage and drainage in cities and towns, and all engineering questions of the day which relate to the public health, are treated in the lecture room, and are again enforced by work in the drawing room. The students are required to design the plumbing for a dwelling or a hotel ; the works for disposal of sewage of a town or country house by irrigation, or other works of a like nature. They are also required to inspect certain towns or buildings in the neighbourhood of Boston or in the City, and to make reports on the same, with suggestions for improvement.

In connection with the sanitary and hydraulic work, a laboratory has been provided for the testing of cements and mortars, and for the study of all questions relating to these materials. The other materials of engineering— their testing, properties, manufacture, &c.—are discussed in the course on the strength of materials, or in the special courses where each material finds its principal use.

Railroad Engineering.—The study of this subject is continued in the fourth year, where the subjects of rolling stock, transportation, management metallic track, rack-rail and other mountain roads, shops, yards, and stations, economic location, &c., are studied in greater detail, and in connection with designs in the drawing room.

An option in *Geodesy, Geology, and Topography,* has lately been arranged which will fit students for special work of this character, and will include extended instruction in Advanced Geodesy and Astronomy, and in the details of Advanced Geological and Topographical field work. A summer school, extending through about six weeks, is included in this course.

In the entire course special stress is laid upon design, as the best means of enforcing the principles laid down in the lecture room, and of giving the student a practical hold upon his subject. Frequent visits are made to works in progress of construction, and to shops and manufactories of different kinds, and to cities like Lowell, Laurence, and Holyoke, where engineering operations are carried on upon a large scale. Besides the regular lectures of the school, special lectures are also given by engineers in the active practice of their profession.

MINING ENGINEERING.

It would be impossible for a student to cover, in a four years' course, all of the studies which are ordinarily included in the term "Mining," which, technically considered, not only comprises the prospecting and exploitation of coal and the ores of the useful metal, but also the metallurgical treatment of these ores. The course of study at the Institute of Technology is, therefore, arranged in three divisions, which have, respectively, a mathematical, a geological, and a chemical basis. The first fits the student for the work of the mining engineer proper—that is, the getting and dressing of ores ; the second division has in view the needs of the mining geologist, whose work is mainly the location and study of ore-deposits; and the third, a chemical division, gives the preparation necessary for the metallurgist.

This triple arrangement of studies is seen at a glance in the following schedule, where the mathematical branches are included in first option, the geological branches in the second option, and the chemical branches in the third option :—

SECOND YEAR.	
FIRST TERM.	**SECOND TERM.**
Chemical Analysis. Physics. German. Analytic Geometry. Surveying. Drawing.	Chemical Analysis. Physics. German. Mineralogy and Blowpipe Analysis. *Options.* 1. Surveying; Differential Calculus. 2. Physical Geography; Microscopical Technology; Chemistry. 3. Surveying; Physical Geography; Chemistry.

THIRD YEAR.	
FIRST TERM.	**SECOND TERM.**
Chemical Analysis. Geology. German. Mining. Physics : Lectures. *Options.* 1. Chemistry; Integral Calculus and Applied Mechanics. 2. Chemistry; Literature; Physical Laboratory; Zoology and Palæontology. 3. Literature; Special Methods; Physical Laboratory; Theoretical Chemistry.	Chemical Analysis and Assaying. German. Mining. Geology. Literature. *Options.* 1. Applied Mechanics. 2. Chemistry; Physical Laboratory; Zoology and Palæontology. 3. Chemistry; Physical Laboratory.

FOURTH YEAR.	
FIRST TERM.	**SECOND TERM.**
Chemical Analysis. Mining Laboratory. Modern History. Ore Dressing and Metallurgy. Memoirs. *Options.* 1. Applied Mechanics. 2. Special Geological Work. 3. Special Metallurgical Work.	Chemical Analysis. Modern History. Metallurgy. Memoirs. *Options.* 1. Mining Laboratory; Motors. 2. Special Geological Work. 3. Mining Laboratory; Motors.

The systematic instruction of this course, which includes practical work in the chemical and physical laboratories, is supplemented by geological field-work and by occasional visits to mines and smelting works. Further, in the fourth year, the students have a course of practical work in the mining laboratory, which contains a very complete plant for the dressing, assaying, amalgamating, and smelting of ores on the large scale.

The equipment of this laboratory is as follows :—

The milling room is supplied with four suites of milling apparatus :—

I. A three-stamp battery, a set of amalgamating-plates, a mercury-saver, a Frue-vanner for concentrating tailings, a settling-tank, and a centrifugal-pump.

II. A Blake challenge crusher, crushing-rolls with automatic sizing screens, a Richards-Coggin separator, a spitzkasten, two Harz-Mountain jigs, an Evans table or rotary-buddle, a settling-tank, and a centrifugal-pump.

III. A set of four amalgamating-pans, 30, 18, 12, and 8 inches in diameter respectively, also a 30-inch settler, and a little automatic kieve for separating mercury from pulp.

IV. A set of three 40-gallon leaching-vessels, a set of four 8-gallon leaching-vessels, and a small dynamo for deposition.

This laboratory contains also the following auxiliary apparatus ; a steam-engine, a Bogardus mill, a Root blower, a Sturtevant dust-fan, drying-tables, and four Morrell agate mortars.

The furnace room contains a water-jacket blast furnace, a copper-refining furnace, a reverberatory lead-smelting or agglomerating furnace, two roasting furnaces, furnaces for cupellation, furnaces for fusion, a blacksmith's forge, a melting kettle, retorts, &c. The assay-room contains ten crucible furnaces, 12×12, all of which are jacketed with iron shalls to insure good draught, stability, and durability ; also two muffles 4×7, one muffle 3×6, four muffles 7×12, one muffle 8×15. These furnaces are all provided with ample flue capacity and abundant draught. This room contains also six pulp-balances, six flux-balances, five button-balances, and desks for fifty students.

The metallurgy of lead, copper, gold, and silver has been selected as the best suited for laboratory illustration. A student is given a quantity of ore weighing, according to its kind, from five hundred to four hundred pounds. This ore he first examines mineralogically and chemically, determining accurately its composition and character. Then arises the discussion of the best metallurgical treatment of such an ore, which necessitates an extensive and careful research, in the appropriate literature of the subject, for available methods. In the actual treatment of the ore, in which he is assisted by one or more of his classmates, he has to test, by assays, every step of the dressing and smelting to ascertain whether the actual composition of the products he obtains agrees with the theory on which he has based his operations.

It is not claimed that this kind of practice is, in any sense, a substitute for experience in regular smelting works. As far as it acquaints students with the machines, tools, and furnaces which he may subsequently have to deal with, it is a not unimportant preparation for such experience. But the primary idea of the work in the mining laboratory is training in habits of observation and investigation. This training can, it is believed, be better learned in the schools than in the works. The business exigencies of work in regular practice, seldom permit a young man to carry on an original and systematic investigation of the nature and conditions of a smelting process. Further, the self-reliance which a student acquires in a year of independent research, with proper oversight and guidance, is worth many years of practice which gives him only manual dexterity, or the knowledge of details of well-known processes. To help a student to acquire this self-reliance, and to teach him how to go about the investigation of a mining or metallurgical problem, is the object sought in the mining laboratory exercises.

ARCHITECTURE.

The course in Architecture is intended to train the students equally for the artistic and the practical side of their work. The instruction in design is modelled as closely as the length of the course will allow, upon that of the Ecole des Beaux-Arts in Paris ; and a large collection of works from the Paris school, comprising examples of everything, from the twenty-four hour sketch designs, to a set of drawings for the Prize of Rome, and including "envois" from the prize students in the school of the French Government at the Villa Medici, are displayed on the walls of the drawing room, or in the portfolio in the library, to serve as standards of execution. No student is admitted to the course who has not, during the first year, the studies of which are common to all the courses, attained more than ordinary rank in drawing. Those who are admitted begin immediately, by lectures and exercises, the study of elementary architectural forms, together with water-sketching and shades and shadows, and, by the beginning of the second term, are ready to begin original design. The first problems are not complex, but careful rendering is insisted upon, and the students are obliged to make constant use of the library, and to store their memory with good detail, not only by the searching and selection necessary for the improvement of their designs, but by systematic exercises in tracing from designated books, and in sketching from casts. Instruction is also given during this year in the history of architecture, and in perspective.

In the third and fourth years of the course, problems in design are given of increasing difficulty, varied with sketch problems, so as to exercise the students upon as wide a range of subjects as possible; and interest and emulation are encouraged by having the problems of each class, as soon as completed, judged, with the award of honourable mentions, by a jury which is very frequently chosen from among the most distinguished architects of Boston, one of whom usually afterwards criticises the designs before all the classes. Together with the instruction in design, practice in water-colour and other sketching, under a professional artist, is continued during the remaining two years, and courses of lectures and exercises are given in mediæval and modern architectural history, in the history and designing of ornament and colour decoration, and in advanced perspective and pen and ink works; and a class in drawing from the life, open to all the students in the department, is held two evenings in each week, under a competent instructor. By the courtesy of the Trustees, the galleries of the Boston Museum of Fine Arts are at all times open to the students of the department, and many of the exercises in sketching, both in black and white, from casts of sculpture and ornament, and in colour from examples of decoration, take place there.

The scientific side of the instruction, during the second year of the course, comprises physics, analytic and descriptive geometry, physical geography, and differential calculus; with a series of lectures on materials of building, and on common constructions. In the third year physics is continued, and lithology and structural geology are taken up. A few weeks are devoted to integral calculus, as an introduction to applied mechanics, which, in the shape of general statics, stresses in frames, strength of materials, and stability of structures, is continued through the remainder of the course. Together with these is given in the third year a series of lectures and exercises in iron construction, working drawings and framing, and stereotomy, together with practice in the simpler operations of surveying, and in the use of transit and level; and in the fourth year lectures are given on planning, on schools, theatres, and churches, on heating and ventilation, and on specifications and contracts, with a long course of lectures and exercises in advanced construction. German is taught in the second and third year, and French in the first year, with advanced French in the fourth year. Each student completes his course with the working out of a thesis problem, illustrating his design by finished drawings, and studying in detail some special point of construction.

CHEMISTRY.

All students who are candidates for a degree attend a course of lectures on Inorganic Chemistry, illustrated by experiments, and perform actual experimental work in the laboratory for general chemistry. This instruction is arranged to meet the needs of engineers, as well as to serve for a preliminary training for those who intend to pursue the study of Chemistry as a profession. Further, it serves as an introduction to the study of science. This course lasts through one year.

The course in Chemistry is designed to prepare students for actual work in connection with manufactures based on chemical principles. The class-room work consists of lectures and recitations on theoretical, analytical, industrial, and organic chemistry. The non-chemical studies, such as mathematics, physics, mineralogy, English history, political economy, and language, are selected mainly with reference to their bearing on chemical work. The student spends a large part of the four years in the laboratories: in the second year the weekly average per student is eighteen hours, in the third year thirty hours, and in the fourth year thirty-five hours. The following scheme shows the subjects taught :—

FIRST YEAR.
Same for all courses.

SECOND YEAR.	
FIRST TERM.	SECOND TERM.
Chemical Analysis.	Chemical Analysis.
Theoretical Chemistry.	Mineralogy and Blowpipe Analysis.
Physics.	Physics.
German.	German.
Political Economy.	Literature.
Analytic Geometry.	*Options.*
	Differential Calculus.
	Physical Geography.
	Microscopical Technology.

THIRD YEAR.	
FIRST TERM.	SECOND TERM.
Chemical Analysis.	Chemical Analysis.
Special Methods.	Theoretical Chemistry.
Industrial Chemistry.	Industrial Chemistry.
Physics: Lectures and Laboratory.	Physical Laboratory.
German.	German.
Literature.	Literature.
Options.	*Options.*
Integral Calculus.	Physics.
Geology.	Geology.
Chemical Analysis.	Sanitary Chemistry.
General Physics (Electricity).	Industrial Chemistry.
Sanitary Chemistry.	

FOURTH YEAR.	
FIRST TERM.	SECOND TERM.
Chemical Analysis.	Organic Chemistry.
Organic Chemistry.	Thesis Work.
Physics.	
Metallurgy.	
Abstracts.	
Options.	
Physics.	
Language.	
Sanitary Chemistry.	
Laboratory Options.	
Analytical Laboratory.	
Organic Laboratory.	
Metallurgical Laboratory.	
Industrial Laboratory.	

While there is a certain prescribed course of study and work in the separate departments of Chemistry, which all regular students must pursue, it will

29

be seen that there is allowed great latitude of choice of subjects in the third
and fourth years.

The instruction in Analytical Chemistry extends through two or more
years. Each student is given a desk in the laboratory, which is open to him
at all times, and he receives personal instruction. Regular students have
analytical work assigned them with particular reference to the course they
are pursuing. This work is so arranged that they obtain experience in a
great variety of methods and processes, and are thus prepared to undertake
any chemical analysis. The more industrious students, and those who work
extra time in the laboratory, have the privilege of supplementing their
regular laboratory course with special work and instruction if they desire it.
Special students may select any branch of analytical work for which they
are qualified.

Particular attention is given to volumetric analysis. A special laboratory
is fitted for this work, and the students are taught to graduate and calibrate
the various instruments of measurement.

As an introduction to original work, each student is required to undertake
a critical examination of some process of analysis, to determine its limits of
accuracy under various conditions, and to make a written report thereon.

The special instruction in the laboratory is supplemented by lectures upon
methods of analysis and manipulation ; and the current chemical literature
in English, French, and German is reviewed by the students, and subsequently discussed in the class-room under the direction of one of the
professors.

Industrial Chemistry is taught by a course of lectures, and by work in the
laboratory of industrial chemistry. A full description of the most important
technical applications of chemistry is given in the lectures. A part of the
lectures are given by persons actively employed in carrying out the processes
which they describe. In the industrial laboratory, the students prepare
chemical products from raw materials. They also undertake the preparation
of pure chemicals. They are taught fractionation and distillation. Particular attention is paid to the preparation of dyes and mordants. A full
course of instruction in bleaching and dyeing is given. It includes scouring,
bleaching of cotton and wool, and the dyeing of yarn and cloth. The
students are taught how to make comparative tests of dye-stuffs, and
qualitative tests to determine the dyes present upon fibres. The students
also become familiar with many of the most useful methods of commercial
analysis. The laboratory instruction is supplemented by frequent excursions to manufacturing establishments where the practical working of
chemical industries can be examined.

The instruction in Organic Chemistry consists of lectures and laboratory work. The theories of organic chemistry are discussed, and the practical applications of these theories described. The work in the laboratory consists of ultimate analysis, preparation of organic products, and original research. Ample opportunities are afforded for the prosecution of investigations in organic chemistry.

The Institute makes available all the facilities of the lecture-rooms and laboratories to teachers who wish to perfect themselves in chemistry, and to persons of maturer years who are engaged in technical pursuits, and who wish to acquire an accurate knowledge of the science. Such persons are admitted without formal examinations, on satisfying the professors in the department that they are competent to pursue to advantage the subjects chosen.

The Kidder Laboratories of Chemistry afford accommodation for five hundred students. The chemical department occupies thirteen laboratories, two lecture rooms, a reading room, balance room, offices, and supply rooms: in all, twenty-two rooms. The laboratory for general chemistry has places for two hundred and eighty-eight students, and is very completely equipped or instruction in elementary chemistry. The analytical laboratory can accommodate one hundred and fifty students, and possesses every convenience for accurate and rapid analytical work. The organic laboratory has places for thirty students. Conveniences are afforded for conducting offensive and dangerous operations in the open air, or in a separate room. The sanitary laboratory contains places for sixteen students. It possesses a very complete outfit for the analysis of air and water, and for the investigation of sanitary problems. The laboratory for industrial chemistry accommodates sixteen students. It contains jacketed kettles, a centrifugal drier, drying chambers, stills, presses, and numerous other pieces of apparatus needed to perform chemical operations upon a considerable scale. In connection with this laboratory is a room devoted to textile colouring, furnished with kettles, water baths, drying room, and various working models of machines used in this branch of applied chemistry.

The reading room of the department contains the chemical library, which now numbers over 2300 volumes, and contains files of all the important chemical periodicals in all languages.

NATURAL HISTORY.

Under this head are included Geology and Biology. These subjects have special claims to a place in any institution of pure and applied science, not only for their pre-eminent importance in developing the faculties of accurate observation, but also for their modern and increasing applications to the arts of the engineer, the physicist, the sanitarian, the chemist, the architect, the teacher, the naturalist, and the specialist.

In the Institute the geological instruction begins in the second year with a half-year's course in physical geography (including dynamical geology) given chiefly by lectures by Professor Niles to students of civil and mining engineering, chemists, natural history students, and general course men. At the same time a course in crystallography and mineralogy is given by Associate-Professor Crosby to the second year miners, chemists, natural history, and (optional) "general" students. This course is determinative in character, and requires the constant handling of specimens. It is accompanied by a shorter course upon practical blowpipe analysis. Those chemists, miners, and natural history men who have followed this course successfully, enter, in the third year, first half, upon a course in structural and chemical geology (in charge of Professor Crosby), in which the constitution, &c., of rocks is thoroughly dealt with. Side by side with this a shorter course in structural geology (intended chiefly as a preparation for historical geology in the second half) is given for the benefit of the civil engineers, architects, intending medical students, and general course men. Historical geology, under Professor Niles, occupies the second half of the third year, and is pursued by the chemists, miners, &c., before mentioned. In all of these geological subjects great stress is laid upon the scrutiny and critical examination of objects or specimens, and to students specially concerned actual problems are appointed, such as the preparation of a geological map of a certain neighbourhood, the explanation of some unusual phenomenon, or some other field work of a practical character.

Palæontology is in charge of Professor Hyatt, who is also Professor of Zoology, and Curator of the Boston Society of Natural History, of which the collections and library, in a building only a hundred feet away, are freely available. During the third year, zoology and palæontology are taught in the same course, the ancient forms shedding light upon the recent, and *vice versa*. Here, also, the student is brought face to face, now with fossils, and now with living species, and is both tempted and compelled to observe and consider.

The strictly biological work begins in the second year, under the supervision of Associate Professor Sedgwick, with a course in general biology for students of natural history, and those intending to follow medicine. The course is made up of lectures, recitations, and laboratory work upon selected forms of plants and animals taken as types. Elementary botany is also given to those desiring it, and is brought home to the student by dissections in the laboratory and excursions in the field. Shorter courses are also given during this term by Professor Sedgwick, in (a) microscopical technology, comprising the use of the microscope, and the preparation, examination, and preservation of specimens, for the special benefit of the chemists, miners, and physicists; and (*b*) in general biology, for students in the general course, who are thus taught to observe closely, and to reason inductively.

During the third year comparative anatomy and embryology are followed by the natural history and preparatory medical students. These courses require prolonged and careful dissections, drawings, and histological work upon the lower animals in the biological laboratory, a large, well-lighted room, supplied with dissecting tables, aquaria, tanks, refrigerator, &c., for comparative anatomy; and with incubators, microtomes, microscopes, models, &c., for embryology. During the fourth year, the special natural history or biological students may follow comparative physiology, doing laboratory work upon the beating heart, measuring the blood pressure, or the phenomena of muscle and nerve, &c., for which myographs, kymographs, sphygneographs, and other modern apparatus, is liberally supplied; or they may continue their zoological, palæontological, or anatomical work still further, pushing it on into the field of original research, under the guidance and with the approval of the professor; or they may take up advanced botany, either morphologically or physiologically; or, if so disposed, may turn once more to geology, and do advanced work in the field or with fossils. During this year, also, the department gives instruction to (*a*) the (sanitary) civil engineers in bacteriology, with microscopic work and practice, especially in the making of "cultures" as applied to water sanitation; and (*b*) a broader course of a similar nature to the "general" students, including lectures upon public health, drainage, water supply, municipal hygiene, &c.

Appendix B.—From Mr T. ALEXANDER, C.E.

APPLIED MECHANICS.

Attended by Students in Civil and Mechanical Engineering, Naval Architecture, Architecture, Chemists, &c.

April, May, and June.—An introductory series of lectures.

Attended by Students of Civil and Mechanical Engineering.

October to March.—A lecture and demonstration course, comprising—analytical and graphical statics—stability of structures—strength of materials—simple stress and strain—lineal elasticity—resilience—loads—bending moments and shearing forces—resistance to bending and shearing—flexure—torsion and compound states of strain—kinematics—kinetics—hydraulics.

CIVIL ENGINEERING.

Third Year Students.

The main subject consists of the professor's lecture and demonstration course on Applied Mechanics, and of a winter lecture course on Civil Engineering. The latter lectures treat of the routine of office and field work, and the requirements and arrangements of plans; of the numerical designing of the details of girders, roofs, bridges, and other work ; and are generally tutorial to the actual business of the drawing office. During this year the students take all their subsidiary subjects enumerated in the prospectus except the winter course in steam, which is taken in the fourth year.

In the drawing office the students make finished designs, from first principles, of simple structures, and of pieces of complex structures. They also study the general arrangement of drawings of civil engineering works from standard designs, and draw copies to modified conditions.

Fourth Year Students.

April, May, June.—Students are required to project a short road, railway, or canal ; and, being divided into companies of three or four, they make the necessary surveys and levels, each student in turn directing the work over one portion.

October to March.—In the drawing office each student makes a plan, a section, and cross sections, of his portion of the above work. He then makes designs, specifications, measurements, and estimates for the necessary bridges and other works ; this occupies three months, and during the remainder of the time designs are made in illustration of the methods treated of in the lecture course. Lectures are delivered on (a) mortars and cements—masonry and brickwork—foundations—tunnelling—road and railway construction ; and (b) the method of the ellipse of stress, and its application to the stability of earthworks and structures under compound stresses in general - strength and stability of bridge trusses –curves of equilibrium for different loadings, and their application to the designing of the suspension bridge the catenary, geostatic and hydrostatic arches and approximate forms—application of hydraulics. The remaining lectures treat of reservoirs, water supply, drainage, rivers and canals, harbours and lighthouses.

Fourth year mining students attend only the portion (a) of the lecture course.

Fifth Year Students.

During their fifth year students visit and reside at the various public works in course of construction by the Government. There they are treated as pupils by the civil engineers in charge, under whose direction they study the construction of the works, make sketches, and take measurements. In some cases a student is entrusted with the charge of a portion of the work. From time to time the students return to the college and make drawings from their sketches, prepare reports, read works prescribed, and pass examinations thereon.

Sixth Year Students.

Part of the sixth year is spent as above, and during the remainder the students are in college preparing their designs and essays for the final diploma examination.

The late Mr William Denny.

At the General Meeting of the Institution, called for 22nd March, at the Rooms, 207 Bath Street, the CHAIRMAN (Mr Dundas) said they were all aware that they met that night in very painful circumstances. Since they last met together as an Institution they had had the misfortune to lose their President—Mr William Denny; and the Council thought it but right and proper that this ordinary meeting of the Institution should be adjourned for a fortnight. They had not had the pleasure of seeing their deceased President in the chair to which he was elected at the close of last session—a chair which he was so well qualified to fill, and to shed lustre upon. He did not think it was necessary at that time to enter upon any lengthened panegyric of the deceased's many excellent qualities, high attainments, and remarkable abilities. They were well known to the whole of the members, and hence they had simply to deplore his loss—a very great loss to the Institution, a very great loss to the scientific world at large, and especially to that of the science of shipbuilding. He might mention that the Council deemed it right that a memoir should be prepared of the late President and printed and issued in the Transactions, along with his portrait. Unfortunately that was the only record of him as President, and which would now have to be accepted in place of the Presidential Address, to which they had been looking forward. Before sitting down, and before adjourning the meeting, he would call upon his friend, their worthy Past President, Professor James Thomson, to move a resolution of condolence and sympathy with the deceased's father and mother, his widow, and other relatives.

Professor JAMES THOMSON said that about twelve months ago he had the honour of proposing Mr William Denny to be their President, and little had he thought then of the sad duty that he was now called on to perform. He would not trust himself at that time to make any biographical sketch, or any remarks on Mr Denny's zealous life, a life in respect of which they were well acquainted. Most of them must well know that he was a man of high ideals and great energy and resolution in bringing his high and good ideas into realisation. He would say no more then, but simply propose the following resolutions :—

" 1. That the Institution of Engineers and Shipbuilders in Scotland desire to convey to Mr and Mrs Peter Denny, and to Mrs William Denny, and to the other members of the family, the assurance of their deepest sympathy with them in their bereavement, and to express their own sense of the great loss which the Institution of Engineers and Shipbuilders in Scotland, as well as the whole public of the West of Scotland have sustained in the lamented death of their late President, Mr William Denny."

" 2. That the Institution request their Secretary to communicate copies of the above resolution to Mr Peter Denny and to Mrs William Denny."

The resolutions were unanimously agreed to, and the meeting was then adjourned till 6th April next.

On Ailsa Craig Lighthouse, Oil Gas Work and Fog-Signal Machinery.

By Mr GILBERT MACINTYRE HUNTER.

Received 19th March ; Read 6th April, 1887.

(SEE PLATES IX. AND X.)

I.—DESCRIPTIVE REMARKS.

AILSA CRAIG may be briefly described as a huge hogbacked rock, rising perpendicularly out of the water for a considerable height. It is situated in Lat. 55°15' 0", N. Long. 5°6' 0" W. in the Firth of Clyde, 10 miles due west from Girvan, and about 12 south-east of Pladda.

Geologically speaking, it is composed of a single rock, and that of a species of trap called syenetic trap of a pale greyish colour, and in some parts, especially on the north side, it is of a reddish hue. On the west and south-west sides it is columnar like basalt, while in others it is amorphous. Numerous trap veins traverse the rock, mostly in a vertical position. The rock, owing to its beautiful appearance and capability of receiving a fine polish, is extensively used for making curling stones. In fact, their fame extends to every quarter of the globe where this national Scottish game is played.

The rock is nearly ¾ mile long, ½ mile broad, is about 2¼ miles in circumference, and covers an area of 220 acres. It is very precipitous on all sides, except the east, where there is a gradual slope, and by which only access can be obtained to the top of the rock. It rises to a height of 1114 feet above the sea level, and can be ascended with a good deal of hard climbing by a circuitous, and in several places very dangerous, path. At the first break in the path,

30

about 390 feet above sea level, stands the ruined castle, a square building 40 or 50 feet high, resembling the "Auld man o' Wick," and several other such buildings throughout Scotland. Very little is known of it, but it is generally believed to have been in connection with Crossraguel Abbey near Maybole, as a religious retreat or abode for refractory monks. On the south wall is seen the remains of an armorial shield bearing three stars cut in relief on a large stone. On the west side of the rock there are the remains of a church called *Ashydoo Church*, which no doubt was used by the monks who resided on the rock. When taking out the excavation for the gasholders, two ancient graves containing human bones were discovered.

Following the path for about 400 feet higher, it descends to "Garra Loch," which is supplied by a powerful spring of excellent water. This loch is of the nature of a "wallee" (in the summer months), it having been probed to a depth of 15 feet without getting bottom.

No trees are to be seen on the rock, save a mere scrub of bourtree near the South Horn, yet the rock at periodic seasons assumes a pleasant touch of nature in a covering of ferns, which vary its aspect—the beautiful white bloom of the sea campion, the blue wild hyacinth, and sea daisies. It is from this rock that the Cassillis family—the chief of the Kennedies—takes the title of Ailsa, but whence the name was derived is unknown. The Gaelic name is *Ailishair-a-chuain*, which means the island of sea fowl or island of sea spray.

Previous to the erection of the work under description, the rock was tenanted only by a tacksman and his workmen, who were periodically engaged in quarrying, fishing, and shooting.

In 1881 Lloyd's Committee petitioned the Northern Lighthouse Commissioners, requesting that fog horns should be placed on the rock to warn ships, and prevent the frequent disasters. In answer to this petition, the Commissioners instructed the Messrs Stevenson, their engineers, to consider the best means of preventing disasters. The engineers, with that view, proposed to erect two fog horns, one on each end of the rock, to be worked from a central station. They also recommended that a light be placed on the south-eastern sides and after some further details were arranged, a party of engineers landed on the rock in July, 1882, and began a survey, a work which could not have been accomplished without incurring great difficulty and danger. The work began in 1883, and was officially inaugurated on 15th July, 1886.

The buildings consist of (1) a retort house and oil stove, with a water tank forming in part the roof, and two gasholders, all in connection with the gas work; (2) the light tower, engine-room, smithy, joiners' shop, and paint store, together with the principal keeper, and three assistants' dwelling houses, each house having all necessary offices; (3) the north and south horn houses, with their respective pipe and cable tracks; and (4) the winding engine, light, railway, and timber jetty, which has a gradient of 1 in 11, so that at low tide access can easily be had from a small boat. All the buildings are, excepting the horn houses and pipe tracks, which are of concrete, of hollow brickwork, the interspace being filled in with concrete. The roofs are flat, and formed of rolled iron joists, covered with buckle plates, the whole of which is then covered with concrete and Val de Travers. The tower rises to a height of 25 feet, terminating in a granite cope, supporting the lantern.

The Firth of Clyde is now provided with a system of distinctive illumination and signalling, such as few, if any other, possesses. At the entrance to Loch Ryan, distant from the Craig 17 miles, there is Corsewall Point Light, alternately white and red; Sanda Isle Light, off the Mull of Kintyre, distant 18 miles; Turnberry Light, on the Ayrshire coast, distant 12 miles; Pladda Light and Fog-Signals, off the southern end of Arran, distant 12 miles; and Holy Isle Green and Red Light, distant 18 miles,—all of which are revolving lights, with the exception of the fixed lights at Pladda and Holy Isle. These lights are all under ordinary circumstances visible from Ailsa Craig.

II.—THE OIL-GAS WORK, AS SUPPLYING THE ILLUMINATING AND MOTIVE POWER.

This section of the work is at once important and interesting in respect without it the "station," as it is officially termed, would be useless as a light and warning, and also as showing the high state of perfection to which the manufacture of oil-gas has reached, the part it will ultimately play in illumination, and in the supplying motive power.

The gas is manufactured from crude mineral oil, or from partially refined shale oil, technically called "blue paraffin oil" by Keith's patent process, consisting of three producers' with four retorts in each producer, in all 12 retorts—each one capable of producing 291·6 cubic feet per hour, or equal to an aggregate of 3,500 cubic feet per hour. One gallon of oil at a bright cherry red heat yields 131·86 cubic feet of gas, or 33,629 cubic feet per ton of oil. Dr Macadam says "that taking weight for weight, oil yields fully three times the volume of gas with at least twice the illuminating power as compared with good average cannel coal." The oil being the extract of shale coal, all ammonia and sulphur escape while it is being distilled, so that these injurious matters being removed, the gas does not require lime or other purification. A highly illuminating gas can be produced when the yield is small, and conversely with a high quantitative yield, *i.e.*, the illuminating value of the gas will be diminished. In illuminating power this gas is about 2½ times greater than Glasgow gas. The cost is about 6s per 1000 cubic feet, and this in illuminating power and consumption is equal to 2 500 cubic feet of Glasgow gas at 3s per 1000 cubic feet, or as 6s against 7s 6d.

Gas when made from mineral oil is of a much higher density and illuminating power than ordinary coal gas, being very rich in light-giving hydrocarbons, it must therefore be very differently treated from coal gas in the manufacture. This is all the more necessary and imperative when economy and efficiency are desired for motive purposes, because the gas being so rich, if it is not properly washed and cooled before passing into the gasholder, will, by constant use, clog up the engine and pipes with tarry matter.

Figs. 1 and 2, Plate IX., are transverse and longitudinal vertical sections of the retort and furnace, the arrows indicating the passage of the flame. It is formed with cast-iron plates, and built with brickwork in fire-clay, all the flues are lined with silicate bricks to withstand the intense heat maintained during distillation. The retorts are placed over the fire, and are protected from the direct action of the flame by a solid flat bottom of brickwork. They are cast-iron pipes, having a flat bottom and resembling in cross section an inverted $\cup$ in order to give a large internal area for distributing the gas, and permitting of their being placed closely together, and fitted at both ends with moveable openings for cleansing purposes. The $\vee$ shaped depression in the top of the retort is intended to augment and complete the distillation of the oil into gas by striking against this depression, and deflecting it towards the sole of the retort. The washer, Fig. 5, Plate IX., is an oblong cast-iron box with a flat top and bottom, a separate pipe leads from each retort into the washer, and on the end of the pipe there is fixed a corrugated "spreader," to distribute the gas over a large area in the water, thus purifying and washing it. The tar collects in the bottom of the box, from which it is run into a cesspool, and ultimately finds its way to the sea. A great loss is thus incurred, as the tar is of good quality, and could easily find a ready sale. After the gas has been thoroughly washed it passes to the coolers, Fig. 6, Plate IX., which consists of two rows of vertical pipes, eight in each row, set in a closed metal base A, with internal divisions b, dipping into water. All the pipes are joined in a special metal box or water seal B, open on the top, and about three-quarters filled with water. The box extends over the whole length of the vertical pipes $A A^1$. Into this water seal two gas-tight boxes C, dip, thus entirely sealing gas-tight the ends of the vertical pipes $A A^1$. By divisions c placed vertically inside these boxes, and the divisions b, at the base A^2, between the alternate pairs of pipes A or A^1, the gas travels up and down the vertical pipes, through the open spaces left at top and bottom, and from one row to the other till it is thoroughly cooled, and from which it passes direct into the gasholder. On the top of the boxes C, there is an outside tray C¹ for water to keep the cover tops cool, having a screw plug in the bottom for emptying them. Any excess of water in the box C, forming the seal at the top passes down the pipes A A¹, into the base A², in which their is an over-flow trap.

In order that the process of manufacture may be clearly understood, the successive stages shall be minutely described. In the oil store there are a series of cisterns, having their respective outlet cocks let into a pipe on the floor line, connecting all the cisterns, and leading into a tank, the top of which is level with the floor. From this tank the oil is forced by a pump to a large overhead tank, from which it flows by gravitation through another pipe to the syphon tubes at the front of each retort, in a thin continuous stream. It then passes down an inclined gutter towards the middle of the retort, where it becomes permanently vaporised, and after passing through the retorts and the ascension pipes, the gas comes to the washer, where all tarry matter or other impurities are deposited. From the washer, the gas next passes through the coolers, and finally into the gas holder. After gas making, if there is soot in the retorts, it indicates that the heat has been maintained at *too high* a temperature; if tarry matter the temperature has been *too low*, while if a *bright cherry red heat*, verging on *white* is maintained, only a small quantity of carbon will be left.

The retort fires are always "banked" and ready for kindling, and in four or five hours thereafter, the retorts are ready for use. The retort doors and all connections are then screwed up, and the oil turned on to each syphon tube, and all stop valves leading to the gas holder, which is to be charged, are opened to allow the gas to pass into it.

In the oil store there are 40 cylinders or drums, containing 95 gallons each; the coal store contains 40 tons; and the service oil tank in the store, contains 280 gallons.

Each gas holder contains 10,574 cubic feet of gas; together they contain 21,148 cubic feet. There are six tiers of plates in the depth of each gas holder, so that each plate represents 1776 cubic feet, while there are 24 rivets in each plate 1-inch centres. As often then as the gasholder rises, the number of rivets multiplied by 74 gives the cubic feet of gas produced.

Each engine consumes 135 cubic feet of gas per hour, thus $135 \times 4 = 540$ cubic feet which is required to work the engines per hour; thus, $\frac{10,574}{540} = 19\frac{1}{2}$ hours supply, being the actual quantity contained in each gas holder.

But the gas is mixed in the meter, immediately *prior* to use with $\frac{1}{3}$ of atmospheric air, therefore, $45 + 135 = 180 \times 4 = 720$ cubic feet of mixed gas, and thus it is increased to the total amount of

$$\frac{10,574 \times 2 + 7,049}{540} = 52 \text{ hours supply},$$

but after deducting the amount required by the tower, which is $17\frac{1}{2} \times 11 = 192$ cubic feet, per average day of eleven hours, thus the *actual* supply is

$$28,197 - 192 = \frac{28,005}{540} = 51 \cdot 861 \text{ hours supply},$$

or 2 days $3\frac{3}{4}$ hours *nett* supply for continuous fog-signalling.

The following results show the variation of the quantity of gas produced from paraffin oil in proportion to the quantity of oil, also the nature of the residual products :—

White heat, $4\frac{1}{2}$ oz. of oil per half minute = 2960 cubic feet of gas produced per hour, cubic feet of gas per gallon of oil, 100. Residue—soot, carbon, and some tar.

Dull heat, 4 oz. of oil per half minute = 2960 cubic feet of gas produced per hour, cubic feet of gas per gallon of oil, 83. Residue—all tar.

Bright red to white heat, 3 oz. of oil per half minute = 2220 cubic feet of gas produced per hour, cubic feet of gas per gallon of oil, 98. Residue—soot in retorts, and tar in washer.

III.—Fog-Signal Machinery.

In the engine-room there are five of Otto's 8 H.P. gas engines, which consume 16·875 cubic feet per H.P. per hour. Four of these engines, or 30 H.P. are required to drive two double cylinder air pumps or compressors with cylinders of 10-inch diameter, and a piston stroke of 20-inch, making 60 strokes, and delivering 218 cubic feet per minute. These pumps are both connected by a 3-inch copper pipe to the north and south air receiver, which are thereby charged with a pressure of 80 lbs. per square inch—together they contain a total of 194 cubic feet compressed air. On the delivery pipe leading to each of the horn houses, there is fixed a diaphragm valve, operated by the compressed air, supplied by a ¼-inch copper pipe, leading from a "pet" valve on the cam gearing. A safety valve and Bourdon Pressure Gauge are attached to each receiver.

The cam gearing consists of a worm-wheel driven by a counter-shaft. It has a projecting flange extending round nearly one-half of the circumference of the wheel, and this flange is made to alternately depress two "pet" valves, one at each side.

There is a patent automatic air and gas meter-mixer for reducing the quality of the gas. The gas as manufactured for the lighthouse is of a very high illuminating power, and therefore not explosive enough for the engines. It consists of an ordinary wet meter, with a divided drum, the one side of which is so arranged that its action is the reverse of the other; thus the gas side of the drum being of a greater capacity than the air side, and the blades of the fan set reversely, the gas side of the drum drives the air side and pumps in the air automatically, and as the drum revolves the mixture is thus formed. In this fashion it is stirred about and passes into a small gas holder or governor on the top of the mixer, which regulates the pressure as it flows into the engine supply pipes. The quantity of gas is registered in the usual manner, prior to mixing. The meter is capable of registering 1500 cubic feet per hour.

On each engine supply pipe there is a safety gas bag or flexible reservoir, with a stop-cock, and also a gas governor or "anti-fluctuator." The side or "pilot" lights of the engines are supplied with gas from the same pipe which supplies the engine cylinders, so that without a governor the engines would be liable to extinguish their own lights. The safety gas bag therefore forms a flexible reservoir from which the engine can supply itself with a charge of gas without setting in motion the whole column of gas in the supply pipe, and also to a large extent, prevents, although not entirely, the fluctuation of the gas pressure in the supply pipe. When it is necessary to keep the pressure in the supply pipe unaffected by the working of the engine, an "anti-fluctuator" is interposed between the safety gas bag and the engine. A balance weight keeps the valve closed, and when the engine draws gas the pressure is consequently reduced and gas escapes from under a diaphragm, then the weight of the long arm of a lever overcomes the balance weight, and slowly opens the valve.

The electric apparatus is operated by a Leclanché Battery contained in a box in the engine-room. It is composed of 20 cells the positive element of which is zinc, and the negative a mixture of carbon and manganese, contained in an inner porous cell. The exciting fluid is a solution of chloride of ammonium or common sal-ammoniac. The solution, containing no corrosive acid, does not act on the zinc plates, except when the current is in circuit. When out of use the battery is switched off. The electromotive force of each cell is about 1·55 volts, and the whole battery gives 31 volts. The resistance of the battery is about 4 ohms, the cable nearly 3 ohms, and the electro-magnet 9 ohms.

The current therefore is $\frac{31}{16} = 1·9375$, or about 2 amperes.

The electric circuit is in use only 6 seconds every $1\frac{1}{2}$ minutes— in order to operate each signal, so that the battery will last a long time without re-charging.

The conducting cables are laid in grooves in a wooden trough, and run up with pitch, and the pipes are of wrought-iron $2\frac{1}{4}$-inch diameter, and all laid in a trench 2-feet deep. The track is carried across the various gullies by means of small span concrete arches, and light lattice girder bridges. The greater part of the north track is provided with a hand-rail on the sea-side.

Figs. 8 and 9, Plate X., show a plan and sectional elevation of the engine-room, and the arrangement of the gas engines and pumps, while Fig. 10, Plate X., shows the automatic and electric apparatus in the horn house. The action of the automatic and electric apparatus in the engine room may be thus described. The compressed air is forced into the north and south air receiver, connected to which are the pipes to the north and south horn houses respectively. The north horn is 1133 yards, while the south is 830 yards distant from the station. The cam-wheel, C, makes one complete revolution every three minutes, or twenty revolutions per hour, corresponding with the automatic gearing in the horn houses, and by means of the projecting flange, D, the "pet" valves are depressed, and thus communicating with the diaphragm valves on the air receiver, and allowing the air to pass to the horn house. The length of time either of these "pet" valves are kept open, regulates the volume of air required to be transmitted after each blast. Co-existent with this automatic apparatus, and as a means of ensuring due and proper action, the electric apparatus is operated by the battery. On the inside of the cam-wheel there is a path in which two studs travel, having a connection with switches or contact-makers—one for making the electric circuit with the north horn, and the other for the south horn. The electric governor is used to break the circuit when the machinery stops. The object of this arrangement is, should it happen that when the engines stopped one of the switches was in contact, the electric circuit would thus be completed, and the armature, F, would not allow the air from the diaphragm to escape, while the automatic apparatus would continue to work until the weight on the pitched chain had completely run down. Unless the cone is in contact with the spring clips, no current will pass, consequently the signals would not be sounded.

The automatic and electric apparatus in the horn houses were designed by Mr Charles Ingrey, and are particularly ingenious and interesting, forming in themselves a study. This is the only instance in which periodic signals have been successfully produced at any great distance from the generating machinery. It will be necessary, however, before proceeding further, to give the distinctive character of each horn. The north signal consists of a single note "Siren," giving a high note of 640 vibrations per second. It is thus composed : a high note of 5 seconds duration, every three minutes. The vane or flyer *s*, is so arranged that five seconds elapse from the time the first projection, *t*, moves the high note valve till the last releases it. The second chain wheel makes one revolution, or four cycles of signals every 12 minutes, or 20 signals per hour. The south signal consists of a double note "Siren," giving successively a high and low note, the former corresponding in character with the high note of the north horn, while the low note equals 280 vibrations per second. It is composed of a high note of two seconds ; silence two seconds; low note two seconds; silence two seconds; high note two seconds ; silence 170 seconds, and equals 180 seconds or three minutes. This vane rotates for 10 seconds, and the intervals of two seconds each are caused by the spaces between the projections *t*. The signals are so timed that the south starts sounding exactly 90 seconds after the north has ceased.

Professor Holmes' "Siren" consists of a revolving slotted drum placed inside a cylinder reversely slotted. In the upper tier of slots, which causes the high note, there are 32 vertical slots, while in the lower tier, causing the low note, there are 14 slots. The air entering through the slots of the cylinder impinges the inclined faces of the slots of the revolving drum, causing it to revolve with great rapidity. The speed of the revolving cylinder is regulated by a centrifugal brake caused by the weight of the blades of the flyer, and in this way regulates the revolutions of the "Siren." The drum is driven at a speed of 20 revolutions per second, the compressed air at a pressure of 80 lbs. per square inch passes freely into the drum and, in passing through the slits to the trumpet, is cut off 32 times in each revolution, thus making the high note of 640 vibrations per second. The low note is cut off 14 times per revolution, thus making a note of 280 vibrations per second. On the pressure gauge in the horn house 10 to 12 lbs. pressure is indicated after each signal, and from 5 to 6 lbs. in the engine-room gauge. The reason of this is due to the pressure remaining in the pipes after the diaphragm is closed, and by the pumps being maintained at their normal pressure of 80 lbs.

The trumpets of the "Sirens" are made of copper, 18-inch in diameter at the mouth, and shaped so that the greater part of the sound is sent seawards. They are pivoted for setting in any direction in azimuth, while the mouths have hinged stoppers or covers which prevent any rain or spray being blown into the "Siren" when in disuse. These covers are blown open by the first blast of the "Siren," and have to be shut by the light-keepers when the signal ceases. The mouth of the north trumpet is distant 17 feet from the face of the rock and directed seawards towards N.E. by N. magnetic, while the mouth of the south trumpet is 25 feet from the rock, and is likewise directed seawards towards S.S.W. magnetic. The signals are sounded during fogs or thick snow showers by day or night, when the atmosphere is so impaired as to obscure the Craig. Under favourable circumstances they are heard at a maximum range of 20 miles.

The automatic apparatus, Fig. 10, Plate X., consists of two chain wheels, b, b', which carry an endless pitched chain, c, and support a wheel and weight, d'. The first chain wheel, b, is prevented from running back by a pawl and ratchet, m, and the second is retained in its normal position by one or other of the projecting pins, f, which rest upon the end of the bolt or catch rod, g.

In connection with the air receiver, A, there is a diaphragm, h, and piston rod, h', to which by means of a conical edge, i, motion is transmitted to the weighted lever, j, so when compressed air is admitted by the pipe, D, beneath the diaphragm, h, the piston rod, h', is forced upwards, when an excess of pressure overcomes the resistance of the lever, j. One end of the lever carries a toothed quadrant, l, which gears into a pinion wheel, l', on the same axis as the chain wheel, b. This pinion, l', is so arranged with a ratchet and pawl that it is free to be rotated in one direction only, but when turned in the opposite direction, it will give motion to the chain wheel, *b*, and by means of the pitched chain, *c*, will raise the suspended weight, *d'*. When the lever, *j*, by its quadrant, *l*, has caused the chain wheel, *b*, to make a partial rotation the pawl, *m*, falls into the ratchet and prevents its being run back by the weight. The quadrant, *l*, carries a curved striker, *n*, so arranged that when the motion of the quadrant is continued it comes in contact with the bell crank lever, *p*, which withdraws the catch, *g*, from contact with the projecting pin, *f*. The weight *d* being now free to descend causes the chain wheel *b'* to be partially rotated until it is arrested by the next projecting pin *f*. Previous to this, however, the catch, *g*, has been returned to its normal position by means of the weight, *q*.

There are a train of pinion wheels, *r*, *r'*. *r''*, *r'''*, connected with the chain wheel, *b'*, which give motion to an adjustable vane or flyer, *s*, and by which the time taken by the weight, *d'*, to descend is regulated. On the circumference of the chain wheel, *b'*, there are a series of projections, *t*, so arranged as to depress or open the "pet" valves, *t'*, *t''*, allowing a supply of compressed air to pass from the receiver, A, through the pipes, *t'''*, to the diaphragm starting valve attached to the "Siren." The whole are so arranged that so long as either of the "pet" valves, *t'*, *t''*, are kept open by a projection, the "Siren" is sounded, but immediately the valve is released the sound ceases, and accordingly as the speed at which the vane or flyer allows the second chain wheel, *b'*, to rotate, and to the length and position of the projections, so is the period of duration or silence of the "Siren."

It will be observed that the diaphragm, *h*, has not only to resist the weighted lever, *j*, but also to overcome a resistance equal to one-half of the weight, *d'*, while the pinion, *l'*, being free to revolve in the reverse direction, and the pawl, *e*, being in gear with the ratchet, *e''*, the weight, *d'*, has no effect in assisting the descent of the lever, *j*. It will be noticed that the diaphragm, *h*—which is of $\frac{1}{4}$-inch india-rubber—will, as it rises, reduce the available area of the cylinder in which the piston, *h'*, works, and the power will be proportionally decreased. To overcome this there is the ball-weighted lever, *r*, which is normally in the position shown. A portion of the weight has first to be overcome by the diaphragm, but the weight becomes reduced as it ascends towards a vertical position, after which instead of offering a resistance it acts as a power in raising the lever, j.

The resistance to the starting valve at 80 lbs. per square inch of pressure is $6 \times 6 \times 80 = \dfrac{2400}{2240} = 1\cdot0715$ ton, the "pet" valve which is less than $\frac{1}{2}$-inch in diameter and requires a force of less than 16 lbs. to operate it.

The air from the receiver, A, passes through the pipe, C, to the electro-magnet, thence through the flutes, G, of the valve spindle to the pipe, D, whence it proceeds to the diaphragm, h, the direction being shown by arrows. The communication is then established and the chain wheels and cams are set in motion. When, however, the electric circuit is broken the armature, F, releases the valve, which rises to the position shown by dotted lines, closing communication between the receiver and the diaphragm, and the surplus air escapes through the pipe, D, to the atmosphere. The weighted lever being now free to fall to its normal position, is ready for the next movement.

As illustrating the progress that has been made in fog-signalling at sea, consider the simple but humane efforts of the pious Abbot of Aberbrothok, who, according to tradition, erected about the thirteenth century a bell on the Inchcape Rock, which he connected with a floating apparatus, so that the winds and sea acted upon it and tolled the bell.

IV.—THE LIGHTHOUSE.

Before proceeding with the description of the lighthouse, it may be well to define the nature and the number of optical apparatus in use, and also the distinctive properties of fixed and revolving lights.

There are three systems of optical apparatus. There are three orders of sea lights, and three orders of harbour lights in each respective system.

The systems are (1) Catoptric, by which the reflection is produced solely by a metallic surface, (2) Dioptric, consisting solely of glass, whether acting by refraction or "total" or internal reflection, and (3) Catadioptric, by a combination of the two previous systems, a metallic with an optical medium. The three orders of sea light; are a First order, having an internal diameter of 1·84 metres = 72·442 inches, and a height of glass of 2·704 metres = 106·457 inches. The Second order, having an internal diameter of 1·4 metre = 55·119 inches, and a height of glass of 2·121 metres = 83·525 inches. The Third order an internal diameter of 1 metre = 39·371 inches, and a height of glass of 1·56 metre = 61·435 inches.

These various orders relate to the size of the optical apparatus, and consequently to their power, thus there may be a First order catoptric light, a Second order catadioptric, and a Third order dioptric light as on Ailsa Craig.

Mr James Chance, M.Inst.C.E.,[*] thus defines fixed and revolving lights :—"For those who have not considered the subject of the *Fresnel* or Dioptric apparatus, it may be well to explain that, according to this system, the source of light is placed in the centre of a structure of rings, or annular segments of glass of such generating sections that all the incident light may be condensed and directed upon the sea. This condensation may take place only in vertical axial planes ; in that case the sea is uniformly illuminated in all directions in azimuth, and the apparatus is termed a fixed light. The sphere of light may, however, be divided into various portions by vertical planes through the centre, and each segment of light may be condensed both vertically and horizontally. The result is a number of separate solid beams ; and in order that they may be seen by the mariner, the apparatus must be made to rotate. This accordingly is called a revolving light."

The Ailsa Craig light, is a Dioptric third order, flashing (" group flashing ") white light, and shows 6 flashes in quick succession for 15 seconds, followed by a period of continuous darkness of about 15 seconds, and makes one complete revolution in 30 seconds. The focal

---

[*] Minutes of Proceedings of the Institution of Civil Engineers, Vol. LVII., p. 168.

plane of the light is 60-feet above H.W. spring tides, giving a range
as seen from the deck of a vessel, under favourable circumstances, of
about 13 nautical miles, and as far round as the rock will permit :
that is, between the bearings of about N.E. by E., round northerly
and westerly to about S. by E., or in other words, down the Firth
of Clyde and on the Ayrshire and Arran coasts, within a sector of
252°.

Fig. 11, Plate IX., gives a section in the plane of the lateral
vertical axis of the apparatus, representing the sections which
generate the successive zones, and these sections are such that all
rays diverging from the principal focus emerge in a horizontal direction. The vertical axis of the burner coincides, of course, with
that of the apparatus. It has a focal distance of 500 millimetres in
the central horizontal plane.

The refracting portion of the apparatus consists of a middle belt,
and 16 zones, 8 of which are above the belt, and 8 similar ones
below it. The whole series subtends in height an angle of 67° at
the focus. There are 13 upper catadioptric prismatic zones, subtending at the focus a vertical angle of 46°, and 4 lower catadioptric
prismatic zones, subtending at the focus a vertical angle of 20°.
Hence, 133° in height, out of 180°, are acted upon by the glass
portion of the apparatus. Each beam is condensed in azimuth from
30° to 10°.

The intensity of the lamp is 60 candle power, and that of the
full power beam in the most illuminated plane is 7632 candles;
the optical aparatus therefore *increases* the initial intensity 127·2
times.

The revolving cupola consists of 12 equal segments or panels,
divided by meridian planes at equal angular intervals, 6 of which
form the flashing compartments, and the other 6 form the dioptric
mirror, 2 panels of which open by hinges, and gives access to the
interior of the apparatus.

The lantern proper consists of framing or astragals for the glass,
an outside sole plate forming a platform for the lightkeepers to
stand upon while cleaning off snow or salt deposited by spray, an

inner service gallery for cleaning the cupola and adjusting the light, and the dome with a cowl fixed in the centre. The framing or astragals are of gun-metal, and somewhat hexagonal in type, with a horizontal astragal dividing the lantern into two tiers. They are securely fixed at the top and bottom, and at the intersection, that they may withstand the heavy gales which are likely to strike against the tower. The glazing is of $\frac{1}{4}$-inch plate glass, having all the arrises carefully rounded off and set in a good bed of putty, and all firmly screwed up. In this way the glass is completely separated from the unyielding material of the lantern.

The ventilation of the lantern forms an important factor in the preservation of a good and efficient light. In a badly ventilated lantern the inside of the panes is continually covered with water by condensation, caused by contact of the ascending current of heated air. The glass, thus obscured, destroys the passage of the rays of light, consequently diminishing the illuminating power of the light. The panes of this lantern are generally free of this evil, although the author observed it on one or two occasions during his stay. There are three ways in this lantern by which it can be lessened, if not avoided altogether : (1) by an upward current from a window near the foot of the tower passing out into the air through the cowl on the dome ; (2) one or other of the doors which communicate with the light room and the balcony can also be used to assist the ventilation, by means of a sliding vertical bolt at the bottom of the door, which is let into one or other of a series of holes in the cope—and the door can thus be kept open at any required angle ; and (3) there are a series of patent ventilators, capable of being shut or opened at pleasure from within the light room, giving a current just on a line with the light keeper's head. The cowl on the dome carries everything up, and allows neither rain nor wind to descend ; even sudden gusts of wind have no effect on the light.

The clockwork consists of two trains of wheels, one for driving the cupola, and the other for an adjustable vane to regulate the speed. The driving weight is suspended by a $\frac{3}{4}$-inch rope, wound round a barrel having friction bearings. A dial plate with an indicating finger is arranged to make one revolution per hour, when an **alarm** bell immediately rings, thus admitting a comparison with the light-room clock. If, therefore, the clockwork should be going too quickly or too slowly, the blades of the vane are accordingly adjusted. The rope barrel makes one revolution every 6·35 minutes, and when fully wound up contains 10 coils of the rope. The driving weight is about 2 cwt., and falls 16 feet 7¾ inches in 57 minutes, when it comes in contact with a tell-tale electric bell, which continues to ring until the weight is lifted off the lever. From the time the electric bell rings, until the weight falls to the floor level, there is one coil of rope on the barrel, which takes 6½ minutes to run off. In this way, the lightkeeper on duty is doubly warned, (1) by the alarm bell on the dial plate, and (2) by the tell-tale electric bell. For a similar purpose, Smeaton placed in his famous Eddystone Lighthouse, a clock which he states,[*] "by a simple contrivance being made to strike a single blow every half-hour, would thereby warn the keepers to snuff the candles."

On the winding barrel of the clockwork there is a self-acting maintaining power, which acts during the winding up. The friction cone rollers—of which there are nine—have each a separate track on the race, to prevent grooving and loss by friction, so that an equal amount of work will be done by each roller, this being the part where the greatest wear occurs. They are consequently made of steel and fitted with great accuracy. There are likewise two tiers of guide-rollers—six in each tier—fixed horizontally and working against the trunk of the cupola framing. The race being the full width of the cupola, there is no outward thrust and the whole apparatus revolves very steadily. By means of a clutch lever the driving spindle can be thrown out of gear.

The lamp produces a steady and uniform flame of 64 millimetres, or 2½ inches, brilliant, pure, and entirely free of smell or "blue centre." A disc of flannel is used for an equal distribution of gas over the valve, which immediately shuts by being pressed into

---

[*] Smeaton's Narrative of Eddystone Lighthouse, p. 170.

the socket, if too much gas enters. When, however, the valve is in its normal position the four outlets are each half open— that is to say, the neck of the valve reaches the centre of the outlets. There are two concentric rings of porcelain, the outside ring containing 50 vertical holes, the inner 30, through which the gas passes. From the gas chamber to these rings, six and three small equidistant tubes are respectively connected. In the centre of the inner ring there is an air pipe to support combustion, while the whole apparatus is surrounded by a perforated gauze frame on the top of which, enveloping the flame, is a flint-glass chimney. The lamp consumes $17\frac{1}{2}$ cubic feet of gas per hour. There is an oil lamp in reserve, in case of the failure of the gas.

The light room is lined with pine boarding, having two doors opening on to the balcony, one on each side, so that the lightkeeper can have access to the balcony in any sort of weather. Communication is established by speaking tubes between the engine-room and lantern, and also between the lantern and the bedroom of each lightkeeper. The watches in the light room are of three hours duration, and are taken by three assistant keepers, the principal keeper is understood not to take watch unless one of the assistants is off on leave of absence, or any other cause. When fogs occur during the night, the keeper on watch in the light room calls the principal and the assistant who has been longest off duty, who together start the engines, and thereafter the keeper on watch returns to the light room, the others remaining on watch in the engine-room. When fogs occur during the day, the principal has to remain on watch, being relieved if it should exceed eight hours.

In conclusion, the author acknowledges his indebtedness to Mr Duncan, the Secretary to the Northern Lighthouse Commissioners, for the permission to inspect the lighthouse and machinery, to Mr David Alan Stevenson, of Messrs T. & D. Stevenson, MM.Inst.C.E., for permission to give this paper, and also to Mr Charles Ingrey, Consulting Engineer to the Pulsometer Engineering Co. (Limited), and to the other contractors, who severally rendered assistance in the preparation of the paper.

Owing to the lateness of the hour, the discussion was adjourned till next meeting.

The discussion of this paper took place on the 10th of May, 1887.

Mr Hunter said that he wished to make a few remarks, to introduce the discussion of his paper, on rapidity of fog-signalling, as they had just had a sad experience in the matter of the wreck of the "Victoria," which was recently lost in the Channel. In regard to this question, *Engineering* of 22nd April, said —"Much undeserved blame has been thrown on to the lighthouse keepers, the idea being that they were remiss in their duty, but they, like the victims of the disaster, reap the fruits of the system on which the Siren is worked by the French authorities, and their inaction, instead of being attributable to a want of industry or watchfulness, arose from the difficulty of getting up steam rapidly." Now the motive fluid was steam, which required about one hour to raise, being a great and unnecessary waste of time. Fogs and snowshowers come on in a few minutes, and much damage could be worked in that time. The Ailsa Craig fog horns could be set in motion and sounded within $3\frac{1}{2}$ minutes from the time the fog appeared, or in the event of the pressure being at zero, it would only take 18 minutes to start signalling. The value and importance of this rapidity cannot be overestimated, when life and property were dependent on a timely warning being given of an approaching danger. It could not be said for the French Lighthouse Authorities that they had not other instances to refer to of rapidity of signalling after fogs come on, when examples have been in existence in this country. He ventured to say that had the Siren been properly and promptly sounded—that is, by means of compressed air, they should not have had to deplore the loss of so many lives, and such a fine ship.

Mr Thomas Burt said with reference to the time stated as required to get up steam—about an hour. Now, he had seen a fire engine the other day get up steam in twelve minutes, so that the time required to get up steam need not be urged against its employ-

ment in working fog-signals. He thought there was a considerable
drawback in the amount and complexity of the machinery on Ailsa
Craig. There were five gas engines, an electrical battery, and two
air compressors, all of which were subject to get out of order; so
that he thought they should try to do something to simplify matters.
All those complex machines were used in very delicate work. The
north horn was 1100 yards and the south horn 830 yards distant
from the station. Was there a man in either horn house, supposing
anything was to go wrong? He believed steam would be more
trustworthy than gas; besides it would keep the attendants warm in
the horn houses. He would not like to say much as to the difficulties arising from this complicated machinery, nor of the propriety of
having more than twice the machinery and power required to do
the work. He did not see that 218 cubic feet of air per minute was
required for the very short blast that was given by the Sirens, for
one of the fog signals only gave one blast of three or four seconds'
duration every three minutes. Again, he thought that was too
long between the blasts, for suppose a steamer was coming towards
the Craig, with the wind, at the rate of 14 knots an hour, before
it heard the second blast it might be in great trouble. He thought
that something might be done to improve these matters, and that
towards the simplification of the machinery was the direction in
which they ought to work.

Mr HUNTER said that while the machinery might not look very
simple, nothing could be more simple in its action and up-keep.
The men on the Craig were all engineers, not lighthouse men. An
experienced principal light-keeper was first tried, but it was found
necessary to displace him by an Engineer Principal. As to the complexity of the machinery, he might say that the south horn had
been in operation for nearly two years and no difficulty had yet
occurred. Then Mr Burt's objection to the infrequency of the fog
signals was not valid, for these signals were given every minute
and a half, as explained in the paper—the one horn sounding 90
seconds after the other had ceased. When the high pitch of the note
of the north signal was considered, making 640 vibrations per

second, it would be seen that it required a great deal of power to drive it.

The CHAIRMAN (Mr Dundas) asked how many men were employed on the Craig, how many were on duty at one time, and how often they made their visits?

Mr HUNTER replied that there were three engineers and the principal keeper, who undertook no lighthouse duty except under special circumstances, but inspected the whole station, took watch in the engine-room while fog signalling, and in the gas work. The assistants had not to go to the horn houses at all. After signalling the cover alone had to be shut, as rain or snow might otherwise be blown into it, but otherwise they required no attention at all. There was only one man on duty in the tower at a time, and when fogs occurred during the night the keeper on watch called the principal and the assistant who had been longest off duty; he then went down and assisted the others in starting the engines, and thereafter returned to the light tower, the others remaining on watch in the engine-room. There were practically only two men on duty while the horns were blowing.

Mr JOHN MAYER asked Mr Hunter if the oil which was used for generating the gas was entirely of the refined character, or if they used some crude oil in the gas making.

Mr DYER inquired to what extent oil gas had been introduced into Scotch lighthouses. He had seen correspondence on the subject between the Irish Light Commissioners and the Trinity House, who after considerable pressure had consented to its introduction into the lighthouses in Ireland. He did not know to what extent it had been introduced into Scottish lighthouses.

Mr HUNTER, in reply, said that both paraffin and crude oil was used on the Craig. There were no purifiers required, and there was scarcely any smell while the gas was burning. Ailsa Craig was the second station under the Scottish Board of Lighthouse Commissioners in which oil had been introduced. The first introduced was at Langness Point on the Isle of Man, five or six years ago, and had been found most satisfactory and economical.

The CHAIRMAN supposed this gas was pretty much the same as that used by the Clyde Lighthouse Trust and the various railway companies for lighting their carriages, and was all manufactured from oil. He proposed a hearty vote of thanks to Mr Hunter for preparing the paper he had presented to them.

The vote of thanks was heartily accorded.

This concluded the business of the Thirtieth Session of the Institution.

Memoir of the late William Denny, F.R.S.E., President of the Institution.

By Mr JOHN WARD,

Member of Council, Institution of Engineers and Shipbuilders in Scotland;
and Member, Institution of Naval Architects, London.

Received and Read April 26th, 1887.

THE very honourable duty assigned to me by the members of Council, to prepare a memoir of our late President, William Denny, is one which I need not hesitate to confess has filled me with inexpressible sadness. To me its necessity means the loss of my best and dearest friend, and the severance of the tie which has bound us together in love and common interests during the greater part of his active life.

The difficulty, too, of such a duty is not less great than is its sadness. I have to speak of no ordinary man. We are too near himself and his work to fully realise either him or it. By-and-by we shall understand him in his true proportions. The shock of his unexpected death is still so sore upon us that we have not yet regained that composure of mind and affection so needful to the realisation of all that he was and did in his short life, and all that we were hoping he would be and do had his life been longer. Some men's characters make impressions on limited areas; they only touch life at a few points, and we can easily enough take the measure of their influence. But the individuality of our late President was so many sided and forceful that no quick survey of his career will give us a full estimate of him. The thought of the undeveloped possibilities that lay within his being adds to our admiration of what he so successfully accomplished. His was a marvellously productive mind, ever ready on any subject, and enforcing and illustrating it with rare gifts of graceful speech. The whole man was projected into every action and scheme, however small it was, and his characteristic earnestness in all undertakings—scientific, educational, social—is that which . raises him beyond an easy comprehension. We shall have to wait for a little ere we can see him and his work in their just lineaments.

He was conspicuously a brave man in life's battle. A much-enduring, courageous soul was in him that ever kept itself bright and buoyant as he held forward towards his high ideal of duty. He was a hero of hard work ; and a terser bit of life history has not often been written out and given in than that which death sealed when he died. He always seemed like one who, with a quick, intuitive perception of the meaning of life and life's obligations and relationships, was eager to do his best, and to do it with manly self-reliance. With a soul that scorned all meanness, duplicity, and sham, and that was ever ardent in praise of all right enthusiasms, he kept a strong grip of principles and facts, and dealt with them not as under a heedless compulsion, but as under the force of truth and justice.

His ready intellectual activity—always so robust and healthy—subordinated to itself all other parts of the man. He thought and spoke clearly as under the pressure of serious convictions, and with his vigorous belief in "the fair brotherhood of men," he moved through the world, ever wishful, with an inspiring unselfishness, to raise all whom he knew to high, pure levels.

These it is, with other things about him, that make him to our memory to-day what he was to our sight and intimate knowledge—a noble, earnest, upright man. Measured by years, his life has been a short one ; but by work done, it exceeds very many that have reached the allotted span.

While my own personal relationship to the late William Denny perhaps justifies me in speaking about him as I now do, it at the same time burdens me with a responsibility all the weightier because of profound personal regrets. For upwards of 15 years I have worked under the spell of his bright and marvellous genius, and for the past ten years a more than common bond of affection has linked us together in his life-work, as well as in the hopes that animated him for what he had planned still to do.

All I shall now attempt is to place on record an account—

brief and general it must needs be—of some of the work done by him for the advancement and welfare of his profession, as well as for those around him, and to express, however inadequately and feebly, in this short memorial notice a tribute of the esteem in which as our President we held him and his labours.   As an Institution we have already expressed our sorrow, and have forwarded a resolution showing our sympathy to his parents and widow.   The Institution of Naval Architects, London, on the motion of their President, Earl Ravensworth, has done the same.   The professional papers, also, of our country have paid fitting tribute to his memory, and the duty that devolves upon us to-night is to accentuate the sentiments so fully, so sympathetically, and so universally expressed.

We may venture to hope that a memoir worthy of our late President will yet be written.   There seems room in the literature of our land for the published record of such a life as his.   It would be easy to show from his utterances and writings that few biographies would be more helpful and inspiring to those beginning their struggle in the battle of life and needful of comfort and sympathy when face to face with the serious issues involved in it.   The purity, the power, the lofty ideals that ever marked the career of William Denny would, if so disclosed, serve as a pattern for those employers of labour who, with large masses of workpeople under their care, are anxious (as he ever was) so to discharge their responsibilities that their servants shall regard themselves as their friends and by the pursuit of a common interest advance a common good.

EDUCATION.

William Denny was born in Dumbarton on 25th May, 1847.   He was early sent to the Academy in his native town, and by the rapidity of his progress in the ordinary branches of education, gave ample promise of his future ability and thoroughness.   At the age of nine he was sent to Jersey, partly for the sake of his health, and partly for fuller training in special subjects of study.   He remained there four years under Dr Carter and the late Rev. A. J. Murray, and at the close of his stay paid, with the latter, a short visit to the East.

On his return he was sent to the Edinburgh High School, where, under the guidance of Mr John Carmichael, a celebrated educationist of his day, he was very thoroughly grounded in a knowledge of the classics. Those who were his schoolfellows then tell with all the vividness of a happy recollection how at the close of the class he would gather them round him and discuss questions that had been raised during the lesson hour, or ventilate projects for the furtherance of his companions' happiness in their games.

At the age of 17 he returned to Dumbarton and entered his father's shipyard as an apprentice. He had resolved to become a shipbuilder, and with a wise prevision for his future usefulness his father made him go through a very real course of training in the several departments necessary to equip him for his position.

HOME INFLUENCES.

Among the influences at work to mould his character for good, there was one which at this time made him very much what he afterwards became. No man had ever more affectionate parents, or loved them with more heroic devotion than he. If it is difficult, as it certainly is, to tabulate the nature and strength of the forces that go to form a man's character, that difficulty becomes all the greater when such home surroundings as were William Denny's have to enter into the calculation. All who know his esteemed father and mother will need no word of mine to corroborate the sincere conviction that he was worthy of all their training and high inspiring example. He inherited much from them both, and, trusting not to any natural powers with which he had been endowed, he ever sought, with the high ideals of conduct set before him by such parents, to live a life of which they would be proud, and in which they would be honoured. It is not my place to say here all that might be said of Mr and Mrs Peter Denny—of their nobility of nature and generous unselfishness of purpose, of their strong sense of goodness and their desire for the happiness of all, the impressiveness of their moral character, the rich warmth of their heart; but I cannot forbear thus alluding to one spring that contributed more than any other thing to make William Denny in all aspects of his being and powers what we knew him to our joy to be. And it is because of this that our sympathy for all who are near and dear to him is so heartfelt and strong.

In January, 1874, he was married to Lelia Mathilda Serena, daughter of the late Leon Serena, of the firm of Galbraith, Pembroke, & Co., of London; his widow and family of four children survive to mourn the loss of a most affectionate husband and father.

APPRENTICESHIP.

The apprenticeship he entered upon in 1864 was characterised by all the thoroughness of the future man. He came and went with the other workmen to work and meals, and made himself from the first one with them in all their interests. Subsequent events in his history as the working head of the firm showed how valuable this discipline was, not for him only, but for all employed in the yard. While he thus gained a knowledge of his business, he also learned to know and handle men. He knew no gulf between master and servant, thanks to his training as a lad in these years of apprenticeship. Understanding, by daily contact and personal friendship with the men, what their hardships and aims and difficulties were, he was all the better able, when he assumed the responsibilities of a partner, to deal with them. The extraordinary tact and generosity and prudence he always showed in his relationships with his workpeople were the fruits of that rich spring-time of learning.

As showing the eagerness with which he desired to qualify himself for his future duties, it may be mentioned that both during and after the close of his apprenticeship, finding himself defective in mathematical attainments, he devoted his early morning and late evening hours to a systematic course of such subjects as were necessary for his technical equipment. Always a hard student, work such as this was a pleasure to him of the highest kind.

IDEAL AS A PARTNER.

In 1868, at the age of 21, he was admitted by his father as a partner of the shipbuilding firm of William Denny & Brothers. The firm derived its name from his uncles and father, by whom it was started in 1844; and for many years, prior to the accession of our late President, carried on business in the yard from which his grandfather, also named William Denny, had turned out pioneer vessels in the era of steam navigation. Later on he became a partner in the engineering firm of Denny & Co.

We obtain a very impressive, hopeful notice of the spirit which animated him on his first admission to partnership, in a speech which he delivered at the annual festival of the workmen, held in December, 1868. It was his first utterance as a master, and he spoke as follows :—"More than four and a half years ago, when I, a boy fresh from school, came among you first, I had little idea of the benefit which my father was conferring on me by putting me in a position where I could get a practical knowledge of the various departments of the work ; I had almost no idea of the many lessons I had to learn. One of these I feel I have learnt, and that, my fellow-workmen, is respect and liking for you. I can look back on these past years without one regret and without one bitter memory. I can thank you without insincerity for the kindness and courtesy that I have met with from all among you. I must also thank those tradesmen under whom my father placed me for the earnest desire they ever showed that I should advance in my profession. Glad, however, as I am that I have been taught a good and honourable trade, I am still more so that I have learned it under many who worked alongside of my uncles, and the greater part of whom will continue, I hope, to work along with my father and myself. Of the future I cannot speak with that confidence which experience gives—my father is my hope as he is yours. I look to him as my pattern and example. I do not know that I shall be able to approach him in business ability, but I shall try to imitate him in looking after your interests and your welfare. In conclusion, I can assure you that your interests and my father's and mine are th same, and that it will ever be my purpose and my desire to make the working men connected with this establishment as happy as I possibly can—God helping."

With this spirit controlling and up-holding him, it is little wonder that with a mind so active, and a love of his profession and work-people so enthusiastic, he should soon spread his views and energies in all channels.

ADMINISTRATIVE WORK.

As the administrative head of a large business he early took an active interest in the Labour Question, and few men in this country have proved better able or tried more successfully to grapple with it in all its bearings.

The problem of piece-work was one of the earliest which engaged his attention, and in which to the very last he took a special interest. In 1871, in conjunction with Mr Ramage, his then General Manager, he organised a system whereby the iron departments of the yard were placed on this footing with satisfactory results. Later on the system of payment by results seemed to him capable of being introduced into other departments with advantage both to the workmen and the firm.

A lecture given before the Dumbarton Philosophical Society in November, 1876, and afterwards published in pamphlet form, entitled "The Worth of Wages," gave rise to much discussion and difference of opinion at the time. At that period his belief was that piece-work rates regulated themselves as time wages did, but larger experience proved that this was not the case. Still, believing the principle to be right, he continued to extend it, the results being effectively controlled so as not to be depressed to too low a point to recoup the workmen for the extra exertion and initiative induced by the very nature of piece-work.

While this subject is one on which much could be said, an extract from almost the last letter he wrote before leaving for South America, alluding to it, expresses his opinion clearly. He wrote:—"The method of piece work is one which cannot either be approved or condemned absolutely, but is dependent upon the spirit and the way in which it is carried out for the verdict which should be passed upon it. It is imperative, in such kinds of piece-work as by their nature cannot be reduced to regular rates, that either the employer should take the responsibility of safe-guarding his workmen's interests, or that the workmen themselves should have an effective control over them. There are, besides, conditions in which even piece-work rates of a general nature may become instruments of very great hardship. I mean instances in which the workers are incapable of effective resistance, and in which employers are either themselves ground down under the force of a competition with which they are unable to cope, or in which, while the employers possess extreme powers of position and capital, they are deficient in any corresponding sense of responsibility to their work people.

"I hope the day is not far distant in which an absentee employer will be looked upon with as much contempt and disapproval as an absentee landlord. If such a healthy public opinion should ever become dominant, it is to be hoped it will be most active in influencing those employers whose works are conducted in great part or wholly upon the piece-work method."

He had a profound belief in conferences for the prevention of strikes and the settlement of differences between masters and men. The recent troubles among our Scotch miners, and the break down of the joint conference of masters and men of that trade, may make the following extracts from a letter written by him, 18 months ago to a large employer of labour, specially interesting. Speaking of his own profession, he wrote :—"I think the aim should be to form an employers' association with the express purpose of selecting and uniting its best men in a permanent conference with an equal number of the men's delegates, the whole to be presided over by some perfectly independent chairman satisfactory to both parties and of known honour and impartiality. . . . .

"The first step is to accustom employers and employed to discuss quietly and fairly their respective interests, and the best way to do this would be to secure every opportunity of bringing even small matters before the combined committee. When I return I shall send you some papers showing how we have been working tentatively towards such objects inside the yard. The worst element in the employers and employed question just now is a very detestable caste feeling making the former look down with an assumed superiority and a good deal of fear and dislike on the latter, the compliment being returned with interest from the latter. You will only eliminate this ugly element by a constant habit of meeting as equals to discuss your interests round the same table. The day that makes our workmen powerless to control and ameliorate their condition will be an ugly one. It would be better to lose our commercial supremacy than face it. I believe the method of permanent conference is the only solution; any way it is worth a trial on the Clyde."

His belief in the goodness of conferences for the settlement of differences was due to the knowledge gained through frequent meeting with delegates from different bodies of his workmen. In 1884 he arranged for a conference to be held with them for the purpose of revising the mode of collecting and dividing the subscriptions made by them to the various institutions and charities which they supported. There were present two delegates from each department in the yard, and the meetings were so pleasant and mutually helpful that they were carried on during 1884, '85, and '86, the labours concluding with a complete discussion and revision of the yard rules.

At the close of the conference in April of last year, he spoke the following memorable words—memorable because they represent the last administrative yard work of his life, and worthy of being read and acted upon far beyond the confines of his native town :—" My own feeling at the termination of these conferences is one of profound regret. To me they have been altogether agreeable and instructive. To come again after several years into close touch with the men in these works has been a pleasure to me. To a certain extent I have been separated from you by business which I have had to conduct outside the works; but I have found, in meeting closely with you once again in these conferences the same straightforwardness and frankness which I found when working among you as an apprentice, and when later I had the outside management under my charge. . . . . .

"Now, gentlemen, the results of our conferences are, first, that we have a workable, just, and effective code of rules. Rules and laws are not, as many people think, valuable in proportion to their severity, for it is to the credit of human nature that wherever a rule or law is too severe it lapses and becomes powerless. Therefore the having just and reasonable rules is the first step towards having them efficient. . . . .

"But apart from all items of improvement, there are three great principles established by these rules and underlying them. The first is that no employer should make a profit by the fines of his workers. I think this principle should have the force of a law. But we have gone further, and have so enlarged the principle for this yard that in all fines we shall be fined along with you, that we shall step down and take our part in the punishment in so far as money can represent it, and that all the money, both your fines and our equivalent to them, shall be returned to you as a body. These principles lift the fines above any suspicion either of profit or vindictiveness.

"The second principle established by these rules is, that as civilised nations now-a-days are not ruled by laws which they have had no hand in making, neither should the workmen of a great public work be ruled by laws which they have had no voice in preparing and approving.

"The third principle is that the fines should be varied, not according to impulse, but according to broad general principles which can be admitted as just, and applied to all future changes in them. I sincerely hope, for the future of the labour question in this country, that these principles may become widely spread. If they do so spread many difficulties at present felt to be grave and doubtful will diminish and disappear. . . . .

"Gentlemen, in the present day we should all extend a little sympathy to each other. We all suffer more or less from the force of a competition which is not confined to our own country, but which presses upon us from all the other civilised countries in the world. The pressure we thus receive we are very much tempted to transmit in its entirety to other people. But this pressure should be divided and fairly borne by all those on whom it comes. No one should seek to transmit the whole of it to those immediately below him, but should keep a portion of the load for himself. He can only ask those who are below him to bear what they are fit to bear, and what it is fair for them to bear.

" Now, in the face of this competition, this commercial war, what should be our policy? It should be one of mutual consideration and mutual forbearance—forbearance on the part of the employer when he has the power to do as he pleases; forbearance on the part of the workman when he has the power to do as he pleases. Indeed, it is only by such mutual consideration and forbearance that the spirit can be cultivated in which we may, as it were by a second nature, live together constantly thinking, in whatever affects us, not only our own thoughts, but the thoughts required for the comfort and happiness of others. In the face of this competition, we should do as an army does in the face of an enemy's charge—close our ranks and draw closer together.

"These conferences, begun by those about your subscriptions, and developing, as they have done, into the series about the rules which we are now concluding, may, I hope, develope still further, and form in the future, as they have done in their short past, an increasingly strong bond of union between us. A bond of union, to be real, implies frankness between the parties who are embraced in it; that they speak to each other plainly about their needs, and by means of much frankness find a way to arrange their difficulties.

" We have not ended our work with these conferences, for over and above the rules which a conference can prepare and sanction, there are other rules which cannot be printed—laws which ought to govern you and us, which cannot be put down on paper. There are laws of justice, fairness, and kindliness which we cannot put between the two boards of a book. These laws must grow, not out of printed paper, but from that human spirit which, when touched within each one of us, can throw up far sweeter growths and flowers than we imagine. We must labour that the spirit of this yard may so strengthen and broaden that injustice from a journeyman to a labourer or apprentice, from a foreman or an under-foreman to anyone in his department, or from a member of this firm to anyone within these gates, shall become a thing impossible and abhorred. That is the spirit we want among us, and we want this yard to have so much of it that injustice may become unknown within its boundaries."

Need I say that the spirit which thus animated the managing partner of Leven Shipyard soon made itself felt upon all who had to do with him. His ideal was high, but not too high for imitation. Those of us who were working under the spell of his earnestness knew that it was no lip service on his part, but a constant practising of what he so fearlessly preached. Nor was it long ere such a spirit bore fruit. For many years past the meetings with the workmen on debatable points almost always left mutually pleasant feelings as a result. As a notable outcome of this, I may mention that during the very busy times of four years ago, his was probably the only shipyard on the Clyde in the affairs of which the Trade Union of Iron Shipbuilders did not seek to interfere.

When the workmen first mooted the establishment of an Accident Fund Society, whereby through voluntary deductions retained from their wages they would be entitled to aliment during the time they were unable to work, he gladly encouraged them in the effort, and ensured its success by giving from his firm towards its funds a sum equal in amount to the total contributions of the workmen. Nor when the Employers' Liability Act came in force was there any change made by the firm.

His own words on this head, spoken at the annual festival of the Friendly Society of Shepherds in December, 1881, of which as of all the other friendly societies in town he was a member, show the true spirit of the man. He said :—" I am betraying no secret when I say your order, in the ceremony of initiation, teaches us a cardinal doctrine, a mutual dependence on each other, as brothers, for help and sympathy in times of distress. That feeling of friendliness and kindness expresses far more than the outer world suspects. It is a great factor frequently lost sight of in the higher ranks of commerce. The feeling of being bound to help each other is one which has been greatest in the world when the world was best, and least when the world was worst. It is the great duty of religion, and is to be taught to the end of time as the great duty of all. It is the expression and reiteration of this which imparts to friendly societies that degree of stability which must sooner or later attract the attention of all who take a deep interest in the more important concerns of our country.

"This aspect of the friendly societies has appealed very directly to all the members of my firm, and influenced them largely in the determination of a most important point. Last year, when addressing a similar gathering, I referred to the Employers' Liability Act. I at that time told them it was a most just act, and one rightly passed. I did not tell them, however, that my firm at the time had under consideration what position they were going to take up with reference to that Act. You have since learned our decision, and in our determination we were largely guided by the knowledge that our workmen were numerously interested in societies like the present.

" We have concluded that men who could form these societies, and who could treat each other in such a friendly way, were worthy to be treated with a similar friendliness by their employers. Therefore my firm decided that they would join no employers' liability assurance company, but take themselves the risk of exceptional cases. We will continue to act as we have done from the beginning, to help our workmen in times of difficulty and distress arising from accident, and take no measures that would in any way divorce our feelings of friendliness from those of our employees."

As a special stimulus to the thinking and initiative powers of his workmen, he in 1880 founded an Awards Scheme, its object being the granting of money awards to any worker employed by the firm, for inventing or introducing a new machine or hand tool; for improving any existing machine or hand tool; for applying any existing machine or hand tool to a new class of work; for discovering or introducing any new method of carrying on or arranging work; or generally, for any change by which the work of the yard is rendered either superior in quality or more economical in cost.

The effect of this scheme has been greatly beneficial, partly in causing improvements to be made, but chiefly in making the workmen of all departments into active thinking and planning beings, instead of mere human flesh and blood machines at so much per day or week. Since its introduction seven years ago, claims have been considered valuable and worthy of award to the number of 196, while rather more than three times this number have been considered altogether. Awards have been granted to the amount of £716. In 1884 he introduced the system of premiums, granting to any workman who had made five successful claims an additional sum equal to the aggregate of the five claims. To show the extent to which the thinking powers of the workmen have been benefited by this stimulus, the total premiums paid in 3¼ years amount to £217, the grand aggregate disbursed cheerfully by the firm for awards and premiums up to date being £933. The benefits of this scheme have been recognised, and the scheme itself adopted, by many employers of labour both in this country and abroad. In every instance the origin has been gracefully recognised.

PROFESSIONAL PAPERS.

Thus far I have taken up a few of the administrative points in the character of our late President, for the purpose of making him more fully understood and appreciated by many who heard or knew him as a scientific shipbuilder only. In turning now to speak of the papers on professional subjects which came from his pen, I shall endeavour to point out some of the practical results which followed from this part of his work.

In 1869, the year after the completion of his apprenticeship, he began publicly to seek to interest others in the problems that had an intense charm and interest for himself. In December of that year he read before the Literary and Philosophical Society of Dumbarton a paper on the "Dimensions of Sea-going Ships." This was followed in January and February of 1870 by a series of four public lectures on the subject of iron shipbuilding, treating—1st, The strains to which ships are subject; 2nd, The means of resisting these strains; 3rd, Strength of iron and its distribution; 4th, Iron workmanship. In order that these might instruct as well as interest, he supplied an incentive to the apprentices of his yard to follow them carefully, by offering prizes for the best essays written by apprentices from notes taken at the lectures.

In 1870 he first carried out the practice upon one of his steamers of trying her at progressive speeds upon the measured mile, and obtaining for her the data for a curve of speed and power; a practice which has since become so widespread, and been recognised as so important in throwing light upon the intricate questions of steamship propulsion. His first paper on this subject was read in Dumbarton in 1871. The further experience and knowledge obtained during the four succeeding years from trials of steamers built in that interval led up to the two papers read by him in 1875 upon the same subject. One of these was read before this Institution, and forms the first of his published papers. It obtained for him the award of the Marine Engineering medal. The other was read before the British Association at Bristol in August of the same year. In 1875, also, he corresponded with the late William Froude, and supplied him with the results of the trials of the "Merkara," "Taupo," and "Hawea," steamers built by his firm, and of the "Arbutus" and "Pachumba," built by Messrs A. & J. Inglis. This correspondence resulted in the production of Mr Froude's well-known and often quoted paper, read before the Institution of Naval Architects in the following year, on "The ratio of indicated to effective horse power as elucidated by Mr Denny's measured mile trials at varied speeds."

As William Denny's earliest published professional paper was delivered to our Institution, so also was his latest. I need do little more than remind you of it. You will remember the vast amount of work which, two years ago, he marshalled before us in it, derived partly from trials of the steamers of his firm and partly from the investigations of the late William Froude and of Mr R. E. Froude ; and you will remember especially the earnestness with which he threw himself into the matter under the sense that it was alike his duty and his privilege to contend for the honour and originality of an investigator in whose steps he was content to be a humble fol- lower, and whom he was proud to regard as both master and friend.

His next two papers, following these read in 1875, were upon Lloyd's numerals. They were both read before the Institution of Naval Architects, one at Glasgow in 1877, and the other in London 1878. They proposed a displacement basis upon which to fix the scantlings, in place of the numerals upon which Lloyd's rules are at present based. One of the objections taken by Mr William John, then chief scientific adviser to Lloyd's Committee to the proposal, was the arbitrary nature of the load line to which a vessel might be immersed, and the consequent uncertain basis which load displacement would present. It is noteworthy, both that this particular objection has disappeared in view of the Tables of Free- board, compiled by the Load Line Committee, and that our past President, Mr Robert Duncan, in the recent Watt lecture at Greenock, reiterated the principle that the load displacement "ought to be the measure of the vessel's strength for her work."

During the course of William Denny's professional career, he twice came into serious conflict with the Board of Trade. The first time was followed by the production of a paper read before this Institution in 1878, drawing attention to what he regarded as the undue inter- ference of the Board in the matter of fitting outlet valves through the sides of a ship, and in demanding drawings under certain circumstances of portions which the builders intended to fit. The second conflict took place in the same year and has proved of far more importance than the first, though there is no published record giving an account of it. It involved the question whether the ton- nage of a structural cellular water-ballast boat should be measured, as was done by the Board, to an imaginary boundary representing what the top of the ceiling would be if the vessel had ordinary floors, or whether it should not rather be measured, as Mr Denny contended it should be, so as to give the internal capacity of the boat, entirely neglecting the space below the inner bottom. The result of this conflict was a decision in Mr Denny's favour, but only after he had instituted an action in the Court of Session did the Board of Trade agree to his contention, and the action was withdrawn.

When the Load Line Committee was appointed by Mr Chamberlain, then President of the Board of Trade, William Denny was one of those invited to act upon it. Their work was of a very laborious nature and lasted for many months, and into it he threw himself with his usual energy ; but though he compiled much, he has not left any separate published account of the Committee's labours.

In 1876 began his first acquaintance with the use of mild steel. The " Taeping " was built in that year of Bessemer steel, for the Irrawaddy Flotilla Company. In 1878 another vessel was built for the same company of Siemens' steel, and this led to the building of the " Rotomahana " in 1879 for the Union Steamship Company of New Zealand, the first merchant ocean-going steamer built of mild steel. The "Rotomahana" was succeeded in the same year by the "Buenos Ayrean," built of the same material. The experience gained at Leven Shipyard from the building of these and other steamers was put by William Denny into a paper read before the Institution of Naval Architects in 1880, and the next year (1881) he followed up the subject by a paper before the Iron and Steel Institute upon the economical advantages of the use of steel.

His next paper, read also in 1881, was the outcome of a desire to make the work of the Science and Art Department more useful to students of Naval Architecture by getting rid of the requirement in the course prescribed by the Department of a knowledge of wooden shipbuilding. This paper led to a deputation from the Institution of Naval Architects to the Government to point out the unsatisfactory nature of this requirement, and the fruit of the deputation's work is that students have now the alternative of wood and iron shipbuilding in the course that they study.

In 1882 he delivered the Watt lecture before the Philosophical Society of Greenock, his subject being " The Speed and Carrying of Screw Steamers." One of the practical results of this lecture was the determination by those responsible for the depth of the James Watt Dock to make it 32 feet at high water. It is needless to say that nowhere else on the Clyde above Greenock is such a depth maintained.

At the meetings of the Institution of Naval Architects the same year he read two papers. One was on launching velocities, and gave results of observations upon the speed with which ships launched in Leven Shipyard pass down the ways. The other was on the treatment of metacentre curves, both transverse and longitudinal, whereby the work done upon a series of known ships can be made of practical value for estimating the stability and trim of proposed new forms.

The sinking of the "Austral" in 1882, and capsizing of the "Daphne" in 1883, made a very deep impression upon him, and, amongst other lessons, he was impressed with one, teaching the need of the greatest possible simplification of matters relating to stability. This feeling led to his conception of cross curves of stability, the construction of which for any ship would simplify the complete treatment of the matter for that ship; and in 1884 he read his paper to the Institution of Naval Architects upon Cross Curves of Stability, the paper taking the form of a description of the methods suggested by two members of his staff for the purpose of carrying out his wishes in the construction of the curves. At this meeting, in speaking in reply to the discussion on his paper, he made use of words expressive of the desire he had ever present before his mind of stimulating those who came under his influence to effort and achievement, taking pleasure in his power of instigation without the shadow of a desire for any of the credit belonging to them. He said—"What I did wish to claim in the paper was the originality of these two methods of Mr Couwenberg and Mr Fellows, in so far as those gentlemen were concerned. I am able to guarantee for them that these methods were worked out independently by them with no suggestion from any other person, that they are the fruit of their own brains, and I may say this to the Institution, because it is an advisable thing that in this

Institution we should not only discuss technical questions, but also questions of *morale* in the treatment of a staff. The principle on which my firm acts with their staff is this : that every particle of credit which belongs to a member of our staff in doing any original work is put to his credit, and carefully kept to his credit. I think he must recognise more and more who has the honour to preside— as I have—over an extremely able staff, that he is presiding over gentlemen in many points of greater capacity than himself, and that he has to deal with them, not as with servants, but as with equals and friends."

In dealing thus with the professional papers that he read, I have said nothing about the discussions in which he took part, especially at the Institution of Naval Architects, upon papers read by others. Into these discussions he often brought a freshness of view, and always a generosity of heart and manner. For those who appeared to be outstripping his own efforts he had nothing but encouragement, and to young or new speakers who had something to bring forward he paid patient attention himself, demanding for them also the patient attention of others. In not a few instances papers read at the Institution were directly prompted by his own personal influence, sometimes exerted on members of his staff, and sometimes upon others. In one case, Mr Heck, a gentleman on Lloyds' staff of surveyors, was encouraged to spend several weeks at Dumbarton in order to carry out his views to the full in the con-struction of a balance which should fairly test his method of dealing with the stability of ships. This was followed by Mr Heck's paper in 1885 before the Institution of Naval Architects.

In the work of the professional and scientific institutions of our country, kindred to this Institution, our late President took a con-stant interest. He was a Member of the Royal Society of Edin-burgh, Member of Council of the Institution of Naval Architects, Member of the Institution of Civil Engineers, Member of the Insti-tution of Mechanical Engineers, and Member of the Iron and Steel Institute. William Denny was ever a deep scientific student and investigator, but he differed from many such in giving his profession and the public, through this and other Institutions, the benefit of his experience. In doing this he never lost the grasp of the yard management, nor did any lack of discipline or want of knowledge of what was being done in all departments result from such scientific and social work.

YARD DEVELOPMENTS.

Finding as he did very special pleasure in carrying on a business which in the hands of his father had become deservedly famous, he was anxious that its capabilities and fame should still further increase under his own management. His native town being entirely dependent upon shipbuilding and its branches, he felt this was a duty essential to the welfare of all the inhabitants.

When opportunity offered of purchasing additional ground, adjoining his works, extensions of a very thorough character were decided upon by his firm. These were commenced in 1881, and embraced an increase of area of 23 acres, viz., from 19 before to 42 acres after the alterations, together with all necessary appliances, including new wet dock, and cranes for tonnage of the very largest class. Special railway, hydraulic, electric, and telephonic systems were laid down, also a number of new departments added. These latter were for the purpose of completing vessels built in all details within the gates, and in more than one instance for the purpose of teaching the daughters of his workmen honourable trades, and helping the town to that extent. There was also added an experimental tank, similar to that owned by the Admiralty, in which many interesting experiments on the hulls of vessels are constantly being carried out. For its construction and appliances he was very greatly indebted to the assistance and advice of Mr R. E. Froude, who is at present continuing the experiments of his late father for the British Admiralty, at Gosport. The importance attached to this branch of investigation is gradually receiving more recognition. This is evidenced partly by Professor Jenkins' recent remarks to his class, suggesting, in the interests of the shipbuilders on the Clyde, the establishment of a similar tank in connection with the Chair of Naval Architecture, and partly by the many requests made to our firm by private individuals and representatives of foreign governments, to have experiments conducted for their benefit and information.

Another matter entered upon at the same time as the extensions of the shipyard, but of wider interest as affecting the other shipbuilders and general welfare of the town, was the improvement and deepening of the River Leven The administration of the Harbour was transferred from the Town Council to a newly constituted Harbour Board, brought into existence by special Act, in the framing and promotion of which William Denny took a large share. He was a member from the first, and rendered valuable assistance in the preparation of the design, and in the construction of the dredger built by Messrs Simons & Co. for the Board in 1885. It was his foresight which caused her capabilities to be largely in excess of present needs, so as to meet future developments.

The thought and worry involved in the carrying out of the yard extensions in all their branches was largely added to by the exceptionally busy times which all shipbuilders experienced five years ago. This was cheerfully faced, but his frame, though strong, was unequal to the continuous pressure. In the autumn of 1882 the strain began to tell upon him, so with his wife he at that time took a prolonged holiday on the Continent.

ILLNESS.

On their return journey, at the end of that year, they were both overtaken by typhoid fever. The attack was a very severe one, but fortunately the symptoms did not fully manifest themselves until they reached this country. Mrs Denny was prostrated when she arrived in London, and was too feeble to travel further. Her husband, however, who on account of pressing duties was wishful to reach home, went on to Dumbarton, but was unable to free himself from the disease. He lay for many weeks in great weakness at his father's house, and it was while recovering from this illness that his own house, Bellfield, with its valuable library, the accumulation of many years, was burned to the ground. This loss grieved him very deeply, though, as was characteristic of him in times of personal misfortune, he did not speak much about it.

It was many months ere he regained strength sufficient to allow him to resume his duties in the yard, the trouble from which he had chiefly to recruit being the specially serious one of an over-worked brain. It was patent to his friends, and he himself felt, that he was not quite the same in bodily health after that long illness. During 1883, after their recovery, he and his wife spent a considerable time travelling on the Continent. Both needed change, and it was hoped that he would return to duty again, having thrown off the effects of the fever and the other trouble.

As manager of the works I was—as I had always been when he was absent—in constant correspondence with him, and many warm-hearted letters which he wrote to me at that time are special treasures in my possession to-day. In one of them, written from Tours, at the end of June of that year, he says :—" Your warm feeling for myself has its partner in my own heart, and I will not readily forget your standing by me in these past weary months. Many a time during them I felt as if my ropes were coiled down for good, and my power of consecutive brain-work gone for good and all. Thank God, this is not so, and the power is returning." In another letter during his prolonged absence, that same year, when speaking of business aims and life, he wrote me as follows :—" I wish success to come to my firm, and to feel I have a share in it ; but beyond this I wish continuity and solidity for the firm, as a thing superior to my own interests or the interests of any single member of it. Life is a very composite item, and no single personal aim can fill or satisfy it, and the outer world least knows its success or failure. The true life, if one could only attain it, is that in which a man marches straight on, purposed to do his work as a duty, to care for others as the noblest pleasure, and to keep his mind unclouded and clear, dominant over his whole sphere of action, by avoiding all intemperance, including the blinding intemperance of the elation of success. But this is an ideal rather to be striven for under many reverses than to be expected. There is no

harm, however, in pitching an ambition high, which involves hurt
to no one."

On his return to duty in September, 1883, it was hoped he
had regained his old vigour, but as soon as he settled down to
work again his system felt the strain very heavy. Still he
laboured on as unweariedly as before, and between then and his
departure for South America in June of last year, some of his best
professional and administrative work was done. To those of us
who knew the exhaustion produced by his constant mental activity,
the fear of a return of the old prostration was often with us.
Knowing as I did his troubles and weaknesses, I often asked him to
take business more leisurely, as did also his medical adviser. His
answer to me more than once was that he believed his life would be
a short one, and his life work must therefore be crowded into a few
years. This prophecy was to prove too true, but how little did we
think the end was so near, and to take place so far from home.

VERSATILITY AND ENTHUSIASM.

Just as the papers on professional work, of which I have given
you an outline, show the enthusiasm and scientific ability of the
man, so did the versatility of his genius come out in relation to all
the different organisations that existed for the good of the people.
Particularly was this so in Dumbarton. All classes of the community there saw that every year the impress of his worth was
growing deeper and deeper on the minds and character of those
with whom he worked. It was noticeable as he grew older, and
especially after his illness of 1883, that he was less sparing of his
strength in such work. As I have just said, he often referred to the
conviction that he would not live to old age, and thus was eager to
compress into a short life what would have been the exacting and
exhausting labour of a long one.

The closing hours of many of his hardest working days were
given to public addresses on questions that were engrossing the
public attention at the time. The sense of his personal responsibility as a man of business and of influence, deepened with his

growth in years. His words more and more carried the tone of a man who was under the force of a wide prophetic outlook on existence. People gathered to hear him when he was announced to speak in public, with the certainty that he had something to say which would stimulate thought in them. The variety of topics that engaged his attention on these occasions may be taken as an indication, not only of the many-sidedness of his genius, but of the wide grasp he had of facts and principles underlying human conduct and effort. We have only to look over the list of his papers read before various societies to see the width of his reading and the extensive field of thought over which he travelled. It was no ordinary man who could speak, and speak impressively and well, on such diversified subjects as "Evolution Theory in Politics, Biography, Literature, and Science," "A Method of Intellectual Culture," "The Worth of Wages," "The Industries of Scotland," "Science and Life," "History of Labour," "Technical Education," "Popularisation of Art," "Public Speaking," "Religion in this Life," and others.

It is quite needless for me to try and specify all the movements with which the name of William Denny will ever be identified. If I associate his name with the following, varied indeed, but all regarded by him as having a useful and commendable purpose, you will see that his life was no idle one, but philanthropic and serviceable in the highest degree :—The Dumbarton Volunteers, from which he retired with rank of Major; the Dumbarton Co-operative Society, the Band of Hope organisation, the Choral Union, the Football and Cricket Club, the School Board, the School of Art, the different Friendly Societies in the town, the Benevolent Society, the White Cross Purity Movement, the National Sabbath School Convention, and others. In the proceedings of the Dumbarton Literary and Philosophical Society he took a very active share. Before he was 26 years of age he had read papers on such subjects as "Dimensions of Sea-going Ships," "Symbolism," "Philosophical and Scientific Inquiry," "Relation of Speed to Power in Steamers," "Condition and Progress of the Arts and Sciences." He had always the enthusiasm of a genuine student. Had he not been a shipbuilder with a famous record in a short life, he would easily gave gained the popular ear as a fresh, inspiring educator.

As showing the form and scope of much of his teaching, we may take a few extracts from one of his most recent addresses, and one which at the time of its delivery created much interest—" The Question of Success." He says :—" Do we not, because we happen to be successful, far too often despise men who are better in heart, nobler, and purer in life than ourselves, because of this blinding of success ? And when we have seized a little truth, when we have made some little pebble off the great shore of truth our own, do not we become blinded and fascinated by it, and forget that it is but one of innumerable pebbles that lie on that great shore ? Our eyes become fixed upon this mesmeric hope—on this single item of success—and we forget the big, wide, many-sided truth that lies outside. When success lifts us above the level at which poverty has no longer a terror for us, don't we far too often forget the companionship of poverty, and, instead of, with St. Francis d'Assissi, taking poverty for our pride, don't we show her to one side, and forget the influence, forget the sorrows, forget all the darkness and shadow that lies on half of the human lives about us ? When we become truly great we shall become like little children. That is not a religious statement, but a scientific truth—that the attitude of all of us before this mysterious universe—this mystery within ourselves and without ourselves—this enormous sense of the duty which raises itself like an awful standard above the individual and the race —if we look at all this, and do not become like little children sooner or later, we are madmen and fools." . . . .

" Do not forget that, taking human life as a whole, what we are engaged in is a great battle. There may be some not taking a proper share in that battle, some who are taking a proper share, and others who are taking a noble share ; but the battle is going on. Humanity is the great fighting line which is advancing with sorrow, with suffering, with pain, and with struggling—with men falling in the ranks dead, and with men suffering and maimed. That is what is going on

36

in the world, so that by struggling and by striving, by sympathy and by sorrow for those who are falling, we should be trained in this advancing principle, this great and limitless aim. Alas, there are men who, instead of being in the fighting line, are lying in luxury behind it, wondering that the world is so hard to live in for some people. Those men forget that if for a day that line were to cease to fight, they and their pampered lives would be swept from the face of this world. These men do not only live and enjoy luxury behind this great struggling line, but actually sometimes draw from it men to serve their luxury. But, thank God! there are other men who go before the fighting line. When we read of human lives sacrificed in the forlorn hope before this fighting line—men who rushed before it, not for the pleasure of commanding, but for the pleasure of being shot down that their bodies may form a bridge across the ditch that the men might advance over it—we have confidence in the beauty of human life, and in the purpose that is guiding us over this miserable struggle; and we feel that although our efforts may be poor and weak, our principles are of a race in which heroes live, and in which we ourselves may have hope." . . . .

"It is a fight, therefore the counsel I give you is this. Be wary of prizing success; be wary not to over-value it; be very wary of despising failure, be patient, and be pitiful to all men; and, above all, looking to the greatness of this life—looking to the infinite possibilities of the majestic aim before us—be submissive, be brave, be undaunted, and be even to the end of your lives—although they end in pain and suffering---perpetually hopeful."

DEATH IN SOUTH AMERICA.

On 16th June, 1886, he left for South America, partly in connection with business of the La Platense Flotilla Company, of which he was a director, but mainly in the hope that his health would be benefited. It was his intention to remain but a short time, but the work demanded more attention than he had anticipated.

His headquarters were Buenos Ayres, and there, as at home, he soon made the impress of his kindly and enthusiastic personality felt.

In a letter received last week from one of the leading Europeans long resident in the country there is this testimony :—" I do not know of any man who after so short a time in the country had made so many friends, and this was evidenced by the large concourse of people, native and foreign, who formed the sad procession to the churchyard. At the grave Dr Irogyen, formerly Prime Minister, spoke a few words in Spanish, testifying to the merits of the departed, adding that it was marvellous how in so little time he had become acquainted with the country and its needs, and had so much won the esteem of the native people with whom he had come in contact."

The separation from home and business was very trying to him. Instead of his sojourn proving in large part a holiday, as it was meant to be, it gathered anxiety and overwork as time went on. This told greatly on his constitution, but to our sorrow no mention of this was allowed to appear in his home letters. He had not given himself sufficient time to regain his strength, and in the work to be done never could spare himself. We now know how anxious his friends were for this excessive mental strain, but none of them dreamt how very soon it was all to be ended by death. The end came, 17th March, 1887.

CONCLUSION.

If as an Institution we have sustained a severe loss, it is surely not too much to say that the shipbuilding industry of our country is all the poorer by reason of his decease. To his family and to his native town the loss is an irreparable one. As I said at the outset, we are thinking to-night of no ordinary man, and I shall not seek to sum up what I have already said about him. You have in your minds the picture of what he was, and many of us feel about ourselves and our duties the power and force of his genius and influence. We are not the only losers by his death. He was more than a shipbuilder and man of business. He was a reforming and elevating force—one whose expansiveness of mind and heart cover the distinctive problems of our time.

Full of honest pride at the high honour you had bestowed on him on the eve of leaving this country, your late President spoke to me of the address he purposed giving us on his return. He had high hopes for the future of our Institution, and under his reign he trusted to its members making it take the pride of place it ought to hold.

What changes have a few months wrought. His voice is never again to stir our breasts and fire our hearts with an enthusiasm akin to his own. Yet his memory will live, and it may be that an outcome from his labours and influence is nearer than we dream. We think of his family's sorrow, his widow and children, his parents, and brothers and sisters. We also feel that he held a large place in our national life. One of Scotland's manliest sons has gone. One of the Clyde's ablest and most thoughtful shipbuilders, one of Dumbarton's best beloved townsmen.

Mr ROBERT DUNCAN, Port-Glasgow, said—

Will you kindly permit me to add a few words on my own behalf, and on behalf of the Institution, to the affectionate tribute to which we have just listened ? If what I am going to say should appear in some sense and in some parts but a repetition of what Mr Ward has so well said already, it will, I trust, at least add the merit of independent testimony to what may be considered the partial eulogy of a more intimate friend.

I knew Mr William Denny well from the time when he began to take an active part in the business of which his father is still, I am glad to say, the distinguished head. My friendly relations with Mr Peter Denny began some five-and-twenty years ago, when he was President and I was secretary of the Scottish Shipbuilders' Association, before its incorporation with this Institution, and when William was still a boy at school. My admiration for the father led me to take a more than ordinary interest in the son, when he began to show his powers and take his place as one of the foremost scientific naval architects of our time. I rejoiced for the honour of his family, of the Clyde, and of Scotland, that we had a man of his promise among us, in the full vigour of early manhood, and had great hopes of his future—now, alas, gone with him in his untimely death.

Of the scientific work done by Mr William Denny during his brief career, it is scarcely possible to speak too highly; but there will be little hesitation here in according the first place to the first paper—if not the earliest of his contributions to the science of naval architecture—which he gave to this Institution on the 23rd of March, 1875, and for which he received the Marine Engineering Medal, by the unanimous award of its members. In that paper "On the Difficulties of Speed Calculation," with the intuition of true genius, he struck out an entirely new path in experimental trials of steam vessels, initiating and advocating the system of progressive trials, with which his name is now universally associated, and which it is not too much to say has influenced all subsequent investigation into the power and speed of steamers. Almost the last—if not the last—of his public work was done on the Load Line Committee, which completed its labours in August, 1885, and on which I had the pleasure of being associated with him for about a year and a half. On that Committee I had many opportunities of seeing the bent of his mind and the scope of his powers, that only such close intercourse could give. His contributions and suggestions to the difficult and complicated investigations which the Load Line Committee had to consider, were of the highest value, and assisted greatly to the harmonious and unanimous decision at which that Committee were eventually enabled to arrive.

But Mr William Denny was more than the scientific naval architect to those who came much in contact with him. He was essentially many-sided; and while science and philosophy were his strong points in all things, he was a first-rate man of business, with a clear head and immense energy—amounting almost to impulsive enthusiasm—and an indomitable will, that allowed no difficulties, physical or intellectual, to stand between him and his purpose. Just and honourable himself, he had a noble scorn of anything that savoured of the opposite, and did not hesitate to express it, but always with a courtesy that disarmed antagonism, while it extorted admiration and respect. Essentially fair and liberal in all his opinions on all subjects, and free from dogmatic bias, even on those he held most strongly, truth and justice were ever with him the first and chief considerations. Whatever did not conform to these requirements found in him an uncompromising opponent, and honest doubt on all others could count on him as a sympathising and generous friend.

When to all these eminent qualifications he added a thoroughly cultured intellect, a richly stored mind, and a large and well-balanced judgment, it was easy to see—even very early in his career—that he was destined, if he lived, to be one of our most distinguished men in whatever walk of life he chose to make his specialty.

Unhappily for us all, but most unhappily for his family and friends, his work is done. He has made his mark on British Naval Architecture. What he would have done further for that, and for his country, those only who knew him best can form any possible estimate. All that is left to us now to do is to place on record, as his countrymen, and as members of this Institution, our estimate of the great loss we have all sustained by his sad and untimely end.

I have now only to add on your behalf that we, as an Institution, accord our warmest thanks to Mr Ward for his exceedingly able and faithful Memoir of Mr William Denny—his friend and partner —our much lamented President. Mr Ward has done his duty well —a most painful duty to him—yet none the less a labour of love, a heartfelt outpouring of the affectionate recollections of many years of the closest business relations and friendships with the subject of his Memoir. And I feel assured that I only express the sympathetic gratitude uppermost in every heart present when I convey to Mr Ward your sincere thanks for the admirable Memoir of our late President, with which he has now favoured us.

10th May, 1887.   Mr ROBERT DUNDAS, C.E., Vice-President,
in the Chair.

The late Mr John Thomson.

The CHAIRMAN (Mr Dundas) said, before beginning the business of this meeting, it was his painful duty to record the awfully sudden death of one of the most prominent members of this Institution. It would be seen by the newspapers that Mr John Thomson had been taken away from them in the prime of life, and in the midst of usefulness. He was for six years a member of the Council, for two years—1881-82—he was a vice-president; in 1883 he was appointed by the Institution as representative on the governing board of the College of Science and Arts; and when that College recently was merged in the Glasgow and West of Scotland Technical College, he was again elected their representative. He had long regularly attended the meetings of this Institution, and not only took a prominent part in its discussions, but contributed, during the 23rd Session, a valuable paper on "The St. Petersburg Water Works, which was unanimously awarded the Institution Medal in the following year. He had no doubt that the memory of this great and good man would long live amongst them. He proposed a resolution of condolence and sympathy with the family, who had thus been so suddenly bereaved.

The resolution was unanimously adopted.

Institution of Engineers and Shipbuilders

IN SCOTLAND

(INCORPORATED).

THIRTIETH SESSION, 1886-87.

MINUTES OF PROCEEDINGS.

——◆◆◆◆◆——

THE FIRST GENERAL MEETING of the THIRTIETH SESSION of the Institution was held in the Hall of the Institution, 207 Bath Street, on Tuesday, the 26th October, 1886, at 8 P.M.

Mr CHARLES P. HOGG, C.E., Vice-President, in the Chair.

The Minute of Annual General Meeting of 27th April, 1886, was read and approved, and signed by the Chairman.

The Railway Engineering Medal, awarded to Mr ANDREW S. BIGGART, C.E., for his Papers "On the Construction of the Forth and Tay Bridges," read Session 1885-86, was presented to the author.

The following Papers were read :—

On " The Safety Governor," by Mr J. W. MACFARLANE; and

On " The Construction and Laying of a Malleable-Iron Water Main for the Spring Valley Water Works, San Francisco, by Mr ROBERT S. MOORE, C.E.

Discussions followed the reading of the Papers and were continued till next General Meeting.

The Chairman announced that the Candidates balloted for had been unanimously elected, the names of these gentlemen being as follows :—

AS MEMBERS :—

Mr PETER DUNN, Civil Engineer, Endrick Villa, Pollokshields.

Mr WALTER MACFARLANE, Iron Founder, 12 Lynedoch Crescent, Glasgow.

Mr R. G. WEBB, Messrs Fleming & Co., Bombay.

AS GRADUATES :—

Mr JAMES BROWN, Study and Employment of Engineering, 89 North Frederick Street Glasgow.

Mr J. B. BUCHANAN, Assistant Civil Engineer, 175 Hope Street, Glasgow.

Mr GILBERT MACINTYRE HUNTER, Assistant Civil Engineer, Caledonian Railway, 3 Germiston Street, Glasgow.

Mr ALEXANDER ROBERTSON, Draughtsman, 111 Kenmure Street, Pollokshields.

———

THE SECOND GENERAL MEETING of the THIRTIETH SESSION of the Institution was held in the Hall of the Institution, 207 Bath Street, on Tuesday, the 23rd November, 1886, at 8 P.M.

Mr ROBERT DUNDAS, C.E., Vice-President, in the Chair.

The Minute of General Meeting of 26th October, 1886, was read and approved, and signed by the Chairman.

The Discussion of Mr J. W. MACFARLANE's Paper on "The Safety Governor" was resumed and terminated, and a vote of thanks awarded to the author.

The Discussion of Mr ROBERT S. MOORE's Paper on "The Construction and Laying of a Malleable-Iron Water Main for the Spring Valley Water Works, San Francisco," was resumed and afterwards adjourned for further discussion at next General Meeting.

A Paper on "Erecting the Superstructure of the Tay Bridge," by Mr ANDREW S. BIGGART, C.E., was read, the discussion being deferred to next General Meeting.

The Chairman announced that the Candidates balloted for had been unanimously elected, the names of these gentlemen being as follows :—

AS MEMBERS :—

Mr PATRICK DOYLE, C.E., F.G.S., Calcutta.

Mr JOHN DUNCAN, Shipbuilder, Ardenclutha, Port-Glasgow.

Mr PATERSON GIFFORD, Mechanical Engineer, 10 Great Wellington Street, Glasgow.

AS GRADUATES :—

Mr THOMAS DICK, Ship-draughtsman, Bowling.

Mr GEORGE FERGUSON DUNCAN, Apprentice Engineer, Ardenclutha, Port-Glasgow.

Mr DANIEL KEMP, Mechanical Draughtsman, 174 Cowcaddens Street, Glasgow.

THE THIRD GENERAL MEETING of the THIRTIETH SESSION of the Institution was held in the Hall of the Institution, 207 Bath Street, on Tuesday, the 21st December, 1886, at 8 P.M.

Mr ROBERT DUNDAS, C.E., Vice-President, in the Chair.

The Minute of General Meeting of 23rd November, 1886, was read and approved, and signed by the Chairman.

The Discussion of Mr ROBERT S. MOORE'S Paper on "The Construction and Laying of a Malleable-Iron Water Main for the Spring Valley Water Works, San Francisco," was resumed and concluded (the author's reply to be forwarded), a vote of thanks being awarded the author.

The Discussion of Mr ANDREW S. BIGGART's Paper on " Erecting the Superstructure of the Tay Bridge," was proceeded with and terminated. A vote of thanks was awarded the author.

Papers on " Improvements of Valves for Steam Engines," by Mr JOHN SPENCE; and on " Steam Engine Counters," by Mr MATTHEW T. BROWN, B.Sc., were read, the discussions being deferred till next General Meeting. Mr G. W. WADE described and illustrated by model his New Friction Grip Lever.

The Chairman announced that the Candidates balloted for had been unanimously elected, the names of these gentlemen being as follows :—

AS LIFE MEMBERS :—

Mr WM. TURNER MacLELLAN, Mechanical Engineer, Clutha Iron Works, Glasgow.

Mr ALEX. AMOS, Railway Contractor, Sydney, New South Wales.

Mr ALEXANDER AMOS, Jun., Mechanical Engineer, Sydney, New South Wales.

AS A MEMBER :—

Mr MATTHEW PAUL, Jun., Engineer, Dumbarton.

AS GRADUATES :—

Mr CHARLES H. GEDDES, Apprentice Engineer, 495 St. Vincent Street, Glasgow.

Mr DONALD KING, Mechanical Draughtsman, 6 Drummond Street, Glasgow.

Mr ROBERT LEE, Jun., Apprentice Engineer, 2 Minard Terrace, Partickhill.

Mr WILLIAM MACGLASHAN, Pupil Civil Engineer, 154 West George Street, Glasgow.

Mr JOHN MACGREGOR, Assistant Civil Engineer, 3 Germiston Street, Glasgow.

Mr ANDREW M'VITAE, Foreman Engineer, 385 Dumbarton Road, Glasgow.

Mr JAMES MACK, Assistant Civil Engineer, 3 Germiston St., Glasgow.

Mr ANDREW T. REID, Apprentice Engineer, 10 Woodside Terrace, Glasgow.

Mr JOHN REID, Student Engineer, 10 Woodside Terrace, Glasgow.

Mr WALTER SCOTT, Ship Draughtsman, 348 Dumbarton Rd., Partick.

Mr JAMES REID SYMINGTON, Assistant Civil Engineer, 204 St. Vincent Street, Glasgow.

THE FOURTH GENERAL MEETING of the THIRTIETH SESSION of the Institution was held in the Hall of the Institution, 207 Bath Street, on Tuesday, the 25th January, 1887, at 8 P.M.

Mr ROBERT DUNDAS, C.E., Vice-President, in the Chair.

The Minute of General Meeting of 21st December, 1886, was read and approved, and signed by the Chairman.

The Discussions of Papers on "Improvements in Valves for Steam Engines," by Mr JOHN SPENCE; and on "Steam Engine Counters," by Mr MATTHEW T. BROWN, B.Sc., were resumed and terminated, votes of thanks being awarded the authors.

A Paper on "The Shafting of Screw Steamers," by Mr HECTOR MacCOLL, was read. A discussion followed, and was continued to next General Meeting.

The Chairman announced that the Candidates balloted for had been unanimously elected, the names of these gentlemen being as follows :—

AS MEMBERS :—

Mr THOMAS ALEXANDER, Civil Engineer, 16 Smith Street, Hillhead.

Mr HENRY MECHAN, Engineer, 17 Fitzroy Place, Glasgow.

Mr ROBERT SIMPSON, B.Sc., Civil and Mining Engineer, 175 Hope St.

Mr PETER ALEXANDER SOMERVAIL, Mechanical Engineer, 85 Burnbank Gardens.

AS GRADUATES :—

Mr EDMUND J. GUMPRECHT, Apprentice Mechanical Engineer, Langdale, Dowanhill Gardens.

Mr DUGALD M'FARLANE, Apprentice Ship Draughtsman, 23 Radnor Street.

Mr DAVID L. M'GEACHEN, Ship Draughtsman, 56 Paterson Street, S.S.

Mr DAVID ROSS TODD, Mechanical Draughtsman, 107 Hope Street.

THE FIFTH GENERAL MEETING of the THIRTIETH SESSION of the Institution was held in the Hall of the Institution, 207 Bath Street, on Tuesday, the 22nd February, 1887, at 8 P.M.

Mr CHARLES P. HOGG, C.E., Vice-President, in the Chair.

The Minute of General Meeting of 25th January, 1887, was read and approved, and signed by the Chairman.

The Discussion of Mr HECTOR MACCOLL's Paper on "The Shafting of Screw Steamers," was resumed, and a vote of thanks awarded Mr MacColl. As the subject was of importance, it was agreed to continue the discussion to the next General Meeting.

As time did not permit of the reading of Mr HENRY DYER's Paper on "The Education of Engineers," it was agreed to hold the Paper as read, so that it might be printed, and the discussion of it taken up at next General Meeting.

Mr RICHARDSON, of North Shields, exhibited and explained his Collision Pad for the Prevention of Loss at Sea. A vote of thanks was awarded Mr Richardson.

The Chairman announced that the Candidates balloted for had been unanimously elected, the names of these gentlemen being as follows:—

AS A MEMBER :—

Mr JOHN HAMILTON HARVEY, Shipbuilder, Benclutha, Port-Glasgow.

AS GRADUATES :—

Mr GEORGE R. HINGELBERG, Apprentice Marine Engine Draughtsman, 48 Hairst Street, Renfrew.

Mr CREE MAITLAND, Apprentice Engineer, 4 Hampton Court Terrace, Renfrew Street, Glasgow.

At the General Meeting of the Institution called for 22nd March, at the Rooms, 207 Bath Street, the CHAIRMAN (Mr Dundas) said they were all aware that they met that night in very painful circumstances. Since they last met together as an Institution they had had the misfortune to lose their President—Mr William Denny ; and the Council thought it but right and proper that this ordinary meeting of the Institution should be adjourned for a fortnight. They had not had the pleasure of seeing their deceased President in the chair to which he was elected at the close of last session—a chair which he was so well qualified to fill, and to shed lustre upon. He did not think it was necessary at that time to enter upon any lengthened panegyric of the deceased's many excellent qualities, high attainments, and remarkable abilities; they were well known to the whole of the members, and hence they had simply to deplore his loss—a very great loss to the Institution, a very great loss to the scientific world at large, and especially to that of the science of shipbuilding. He might mention that the Council deemed it right that a memoir should be prepared of the late President and printed and issued in the Transactions, along with his portrait. Unfortunately that was the only record of him as President, and which would now have to be accepted in place of the Presidential Address, to which they had been looking forward. Before sitting down, and before adjourning the meeting, he would call upon his friend, their worthy Past President, Professor James Thomson, to move a resolution of condolence and sympathy with the deceased's father and mother, his widow, and other relatives.

Professor JAMES THOMSON said that about twelve months ago he had the honour of proposing Mr William Denny to be their President, and little had he thought then of the sad duty that he was now called on to perform. He would not trust himself at that time to make any biographical sketch, or any remarks on Mr Denny's zealous life, a life in respect of which they were well acquainted Most of them must well know that he was a man of high ideals and great energy and resolution in bringing his high and good ideas into realisation. He would say no more then, but simply propose the following resolutions:—"*First.* That the Institution of Engineers and Shipbuilders in Scotland desire to convey to Mr and Mrs Peter Denny, and to Mrs William Denny and to the other members of the family, the assurance of their deepest sympathy with them in their bereavement, and to express their own sense of the great loss which the Institution of Engineers and Shipbuilders in Scotland, as well as the whole public of the West of Scotland have sustained in the lamented death of their late President, Mr William Denny." "*Second.* That the Institution request their Secretary to communicate copies of the above resolution to Mr Peter Denny and to Mrs William Denny."

The resolutions were unanimously agreed to, and the meeting was then adjourned for a fortnight.

THE SIXTH GENERAL MEETING of the THIRTIETH SESSION of the Institution, adjourned from 22nd March, was held in the Hall of the

Institution, 207 Bath Street, on Wednesday, the 6th April, 1887, at 8 P.M.

Mr ROBERT DUNDAS, C.E., Vice-President, in the Chair.

The Minutes of General Meeting of 22nd February, 1887, and of Meeting called for 22nd March, were read and approved, and signed by the Chairman.

The discussion of Mr HECTOR MacCOLL's Paper on "The Shafting of Screw Steamers," was terminated, a vote of thanks being awarded Mr MacColl.

The discussion of Mr HENRY DYER'S Paper on the "Education of Engineers" was resumed, Mr T. D. WEIR moving: "That in prospect of immediate legislation affecting the Scottish Universities, the Council be authorised to take action with a view to securing the direct representation of this Institution on the governing body of Glasgow University."

The motion was seconded by Mr P. S. HYSLOP.

The discussion was continued to next General Meeting. It was agreed that, to facilitate the business of the Annual Meeting, Members taking further part in this discussion should send their remarks to the Secretary prior to date of Meeting.

Mr GILBERT MACINTYRE HUNTER read his Paper on "The Ailsa Craig Lighthouse, Oil Gaswork, and Fog-Signal Machinery," the discussion of which was carried to next General Meeting.

Messrs JOHN TURNBULL, Jun., and PETER STEWART were unanimously chosen to audit the Treasurer's Annual Financial Accounts.

The Chairman announced that the Candidates balloted for had been unanimously elected, the names of these gentlemen being as follows:—

AS MEMBERS :—

Mr J. T. BAXTER, Civil Engineer, 9 Brighton Terrace, Govan.

Mr D. J. RUSSELL DUNCAN, Civil Engineer, 10 Airlie Gardens, Kensington, London, W.

Mr EDWARD MACKAY, Mechanical Engineer, 8 George Square, Greenock.

38

Mr WILLIAM NISH, Mechanical Engineer, 15 Govan Road, Glasgow.
Mr THEODOR J. PORETCHKIN, Marine Engineer, Russian Imp. Navy.
Mr JAMES WHITEHEAD, Mechanical Engineer, 71 Scott St., Glasgow.

AS GRADUATES :—

Mr JOHN C. PRESTON, Mechanical Draughtsman, 27 Town Hall, Brisbane, Queensland.
Mr JOSEPH WILLIAM RUSSELL, Apprentice Engineer, 461 St. Vincent Street, Glasgow.

———

THE THIRTIETH ANNUAL GENERAL MEETING of the INSTITUTION was held in the Hall of the Institution, 207 Bath Street, on Tuesday, the 26th April, 1887, at 8 P.M.

Mr C. P. HOGG, Vice-President, in the Chair.

The Minute of General Meeting of 22nd March, 1887, was read and approved and signed by the Chairman.

The Annual Financial Statement was submitted by the Treasurer, and adopted, a vote of thanks being awarded the Auditors of the Accounts.

The Report by Library Committee was held as read. The Report to be printed in Annual Volume.

The CHAIRMAN intimated that the Council had agreed that no medals should be given for session 1885-86, but that premiums of books might be awarded.

On the motion of Mr J. M. GALE, seconded by Professor JAMES THOMSON, a premium of books was awarded Mr Henry Dyer for his Paper on "The Present State of the Theory of the Steam Engine, and some of its Bearings on Current Marine Engineering Practice."

On the motion of Mr GEORGE RUSSELL, seconded by Mr A. C. KIRK, a premium of books was awarded Mr Andrew S. Biggart for his Paper on "The Forth Bridge Great Caissons, their Structure, Building, and Founding."

The Election of Office-Bearers then took place :—

Mr ALEXANDER CARNEGIE KIRK, proposed by Mr Robert Duncan, and seconded by Mr Richard Ramage, was unanimously elected President.

Mr DAVID JOHN DUNLOP, proposed by Mr A. C. Kirk, and seconded by Mr John Ward, was unanimously elected a Vice-President; Mr HENRY DYER, proposed by Professor James Thomson, and seconded by Mr Thomas D. Weir, was unanimously elected a Vice-President; and, by a majority of votes, the following gentlemen were elected as Councillors :—Mr JOHN WARD, Mr JAMES MOLLISON, Mr JOHN TURNBULL, Jun., Mr MATTHEW HOLMES, Mr JAMES CALDWELL, and Mr THOMAS KENNEDY.

Mr JAS. M. GALE was unanimously re-elected Treasurer.

Mr JOHN WARD read a Memoir of the late Mr William Denny, President of the Institution.

Mr ROBERT DUNCAN, after some sympathetic remarks, proposed a vote of thanks to Mr Ward for the excellent Memoir of the President, which he had just read.

The vote of thanks was heartily accorded.

It was agreed that, as the evening was advanced, the discussion of Mr Dyer's and of Mr Hunter's Papers should be adjourned till 10th May, when a Special Meeting should be held for this purpose.

The CHAIRMAN announced that the Candidates balloted for had been unanimously elected, the names of these gentlemen being as follows :—

AS A MEMBER :—

Mr ARTHUR WATSON THOMSON, Civil Engineer, 5 Hagg Crescent, Johnstone.

AS A GRADUATE :—

Mr THOMAS BELL, Engineering Draughtsman, Little Park Cottage, Yoker.

A SPECIAL GENERAL MEETING of the INSTITUTION was held in the Hall, 207 Bath Street, on Tuesday, the 10th May, 1887, at 8 P.M.

Mr ROBERT DUNDAS, C.E., Vice-President, in the Chair.

The CHAIRMAN before commencing the business of the Meeting, referred to the loss which the Institution had sustained through the sudden death of Mr John Thomson, who had for so long taken an active and important part in the work of the Institution, both as a Member of Council and otherwise. He proposed that the meeting should pass a resolution of condolence and sympathy with the family in their trying bereavement.

The resolution was unanimously adopted.

The Minute of Annual General Meeting, of 26th April, was read and approved, and signed by the Chairman.

The continued discussion of Mr Henry Dyer's Paper on "The Education of Engineers" was then proceeded with and terminated, a vote of thanks being awarded Mr Dyer for his Paper.

Mr Weir's motion, as moved and seconded, at General Meeting of 6th April, was then put to the Meeting, and unanimously agreed to.

The discussion of Mr Gilbert M'Intyre Hunter's Paper on "The Ailsa Craig Lighthouse, Oil, Gaswork, and Fog Signal Machinery" was then proceeded with and terminated, and a vote of thanks awarded to Mr Hunter for his Paper.

A Model of Mr Bremme's New Triple Expansion Valve Gear was exhibited and described.

Mr W. Renny Watson exhibited specimens of the handicraft of the Students of the Massachusetts Institute of Technology.

TREASURER'S STATEMENT—1886-87.

GENERAL FUND.

Dr.					Cr.			
To Balance in Union Bank at close of Session 1885-86,			£174 19 1	By Amount paid Treasurer of House Committee as Institution's proportion of Expenditure, for Session 1886-87,			£170	0 0
Subscriptions received :				Printing, ...			162	1 9
Session 1886-87,	£628 10 0			Lithography,			89	11 11
Arrears of Previous Sessions,	40 0 0			Graduate Section Medal, Session 1885-86,			1	7 6
	£668 10 0			Salary to Secretary, ...			125	0 0
Deduct Entry Money transferred to Building Fund, ..	6 10 0			Commission Collection of Arrears of Subscriptions, viz. :—				
		662 0 0		For Session 1886-87,	... £386 0 0			
Sales of Transactions, ...		11 0 1		For Previous Sessions,	... 40 0 0			
Bank Interest, ...		0 19 8			£426 0 0 at 5 %		21	6 0
				Postages, and Delivery of Annual Volumes,...			45	19 0
				Stationery, &c., ...			13	14 11
				New Books for Library, ... :			19	19 9
				Binding Periodicals in Library, ... :			4	12 0
				Cash to New Buildings Account to meet Interest on Loan, from Medal Funds, ... :			16	0 0
				Petty Cash, ...			1	16 6
				Balance in Union Bank, ... :			177	9 6
			£848 18 10				£848	18 10

MARINE ENGINEERING MEDAL FUND.

Dr.		£	s.	d.	Cr.		£	s.	d.
To Balance in Union Bank at close of Session 1885-86,	...	44	11	1	By Balance in Union Bank,	...	54	19	10
Interest on Capital lent to New Buildings Account,	...	10	0	0					
Bank Interest,	...	0	8	9					
		£54	19	10			£54	19	10

RAILWAY ENGINEERING MEDAL FUND.

Dr.		£	s.	d.	Cr.		£	s.	d.
To Balance in Union Bank at close of Session 1885-86,	...	40	9	9	By Medal, Session 1885-86,	...	10	0	0
Interest on Capital lent to New Buildings Account,	...	6	0	0	Balance in Union Bank,	...	36	17	8
Bank Interest,	...	0	7	11					
		£46	17	8			£46	17	8

GRADUATE MEDAL FUND.

Dr.		£	s.	d.	Cr.		£	s.	d.
To Balance in Union Bank at close of Session 1885-86,	...	21	19	8	By Balance in Union Bank,	...	22	9	1
Bank Interest,	...	0	9	5					
		£22	9	1			£22	9	1

BUILDING FUND.

Dr.				Cr.		
To Balance in Union Bank at close of Session 1885-86,	£293	8	10	By Balance in Union Bank, ...	£322	18	0
Entry Money, ...	6	10	0			
One Life Member at £20, ...	20	0	0			
Bank Interest, ...	2	19	2			
	£322	18	0		£322	18	0

NEW BUILDINGS ACCOUNT.

Dr.

To Capital to meet Cost of New Buildings, viz.:—
From General Fund, ... £542 15 7
" Marine Engineering Medal Fund, ... 351 11 2
" Railway Engineering Medal Fund, ... 213 13 3
" Building Fund, ... 939 8 1
£2,047 8 1
Cash received from General Fund to meet Interest on Loans, ... 16 0 0
£2,063 8 1

Cr.

By Paid on New Buildings, ... £2,047 8 1
Interest on Loans, viz.:—
To Marine Engineering Medal Fund, £10 0 0
" Railway Engineering Medal Fund, 6 0 0
16 0 0
£2,063 8 1

GLASGOW, *15th April, 1887.*—WE have examined the foregoing Annual Financial Statement of Treasurer, the Accounts of the Marine and Railway Engineering Medal Funds, the Graduate Medal Funds, the Building Fund, and the New Buildings Account, and find the same duly vouched and correct, the Amounts in Bank being as stated.

(Signed) JOHN TURNBULL, JR., } Auditors.
PETER STEWART,

SUBSCRIPTION ACCOUNT.

Dr				
To Subscriptions due as per Roll:—				
Arrears due at close of last Session, ...	£135	0	0	
Deduct Irrecoverable, ...	57	10	0	
	£77	10	0	
Add elected at Annual General Meeting, April, 1886, ...	3	10	0	
				£81 0 0
SESSION 1886-87:—				
381 Members at £1 10 0 —£571 10 0				
5 New Members „ 2 10 0 — 12 10 0				
4 „ „ 2 0 0 — 8 0 0				
9 „ „ 1 10 0 — 13 10 0				
34 Associates „ 1 0 0 — 34 0 0				
185 Graduates „ 0 10 0 — 92 10 0				
				732 0 0
				£813 0 0

Cr.				
By Subscriptions received, as per Cash Book, viz.:—				
Arrears of Sessions previous to Session 1886-87, ...				£40 0 0
SESSION 1886-87:—				
331 Members at £1 10 0 —£496 10 0				
5 New Members „ 2 10 0 — 12 10 0				
3 „ „ 2 0 0 — 6 0 0				
3 „ „ 1 10 0 — 4 10 0				
31 Associates „ 1 0 0 — 31 0 0				
156 Graduates „ 0 10 0 — 78 0 0				
				628 10 0
Arrears due for Session 1886-87, ... £103 10 0				
Arrears due for previous Sessions, 41 0 0				
				144 10 0
				£813 0 0

BANK ACCOUNT.

Dr.				
To Balances at close of Session 1885-86:—				
General Fund, ...	£174	19	1	
Marine Engineering Medal Fund, ...	44	11	1	
Railway Engineering Medal Fund, ...	40	9	9	
Graduate Medal Fund, ...	21	19	8	
Building Fund, ...	293	8	10	
Amounts lodged, Session 1886-87, ...	518	9	10	
Interest, Session 1886-87, ...	5	4	11	
	£1,099	3	2	

Cr.			
By Amounts Drawn, Session 1886-87, ...	£484	9	1
Balances in Union Bank, ...	614	14	1
	£1,099	3	2

CAPITAL ACCOUNT.

GENERAL FUND.

		£	s.	d.	£	s.	d.
Loan to New Buildings Account,	... ...	542	15	7			
Cash in Union Bank,	... ...	177	9	6	720	5	1

MARINE ENGINEERING MEDAL FUND.

		£	s.	d.	£	s.	d.
Loan to New Buildings Account,	... ...	351	11	2			
Cash in Union Bank,	... ...	54	19	10	406	11	0

RAILWAY ENGINEERING MEDAL FUND.

		£	s.	d.	£	s.	d.
Loan to New Buildings Account,	... ...	213	13	3			
Cash in Union Bank,	... ...	36	17	8	250	10	11

GRADUATE MEDAL FUND.

		£	s.	d.
Cash in Union Bank,	... ...	22	9	1

BUILDING FUND.

		£	s.	d.	£	s.	d.
Amount to New Buildings Account,	... ...	939	8	1			
Cash in Union Bank,	... ...	322	18	0	1,262	6	1

ARREARS OF SUBSCRIPTIONS.

		£	s.	d.	£	s.	d.
Arrears due for Session 1886-87,	... ...	103	10	0			
Do. previous Sessions,	... ...	41	0	0	144	10	0
					£2,806	12	2

DR.   HOUSE EXPENDITURE ACCOUNT. (ABSTRACT 1886-87.)   CR.

			£ s. d.
To Rents for Letting Rooms, ...			£94 12 7
Amounts Received by Treasurer to meet Expenses, viz.:—			
From Institution of Engineers and Shipbuilders, ...	£170 0 0		
From Philosophical Society, ...	131 17 4½		301 17 4½
			£396 9 11½

		£ s. d.
By Balance due Treasurer, ...		£21 19 1½
Interest on Bond, ...		130 10 0
Salary to Curator, ...		95 0 0
Salary of Attendant at Library, Cleaning, &c., ...		40 5 9
Taxes, ...		28 8 5
Feu-duty, ...		0 18 0
Gas Rates, ...		13 15 1
Water Rates, ...		6 13 4
Coals, ...		7 18 0
Insurance, ...		7 15 0
Repairs, ...		22 12 3
Furnishings, ...		4 11 6
Balance in Treasurer's hands, ...		16 8 6
		£396 9 11½

The Account of the House Committee is kept by Mr John Mann, C.A., Treasurer to the Committee, and is periodically audited by the Auditors appointed by the Institution and the Philosophical Society.

W. J. MILLAR, *Secretary to House Committee.*

DECEASED MEMBERS.

DURING the course of the Session 1886-87 the Institution has lost
several gentlemen who have taken an active part in the management
of its affairs, several of whom have from time to time contributed to
the papers read and the discussions arising.

These gentlemen are — Sir JOSEPH WHITWORTH, *Honorary
Member;* Mr WILLIAM DENNY, Mr JOHN FERGUSON, Mr JAMES
HUNTER, Mr WILLIAM M'ONIE, Jr., Mr JOHN THOMSON, and Mr
ROBERT WYLIE, *Members;* Mr IVAN MAVOR, *Graduate.*

Sir JOSEPH WHITWORTH, Bart., was elected an Honorary Member
of the Institution in 1863. After receiving a good education he
went to learn the cotton spinning business, and, in the year
1821, when he was about eighteen years of age, he went to
Manchester, and entered the employment of Messrs Craigton & Co.
there, where he seems to have remained about four years, after
which he went to London, where he worked in Maudslay's and other
works, and returned to Manchester in 1833, where he commenced
business as a tool maker. During the next twenty years he was
busily engaged improving existing machinery and devising new
appliances, at which time also he perfected the methods of measurement through his true planes and delicate measuring instruments,
uniformity in the standard for screw threads, &c., by which the
name of Whitworth became known far and near. The design and
construction of guns followed later on, together with his fluid
compressed steel, now used extensively for the shafting of large
steamships. The name of Whitworth, besides being known so
widely amongst the inventions of the last fifty years, commands the
grateful respect, especially of young engineers, from the munificent
bequest which he made to them in 1869, by the foundation, at an
expense of £100,000, of the " Whitworth Scholarship," which has

for so many years proved a valuable stimulus to the progress of · mechanical engineering in its highest aspects.

Mr WILLIAM DENNY was elected a Member of the Institution in 1873, was a Member of Council during Sessions 1876-77, 1877-78, and again during Sessions 1882-83 and 1883-84, and at Annual General Meeting held on 27th April, 1886, was unanimously elected President of the Institution, an office which, had he been spared to occupy, would have been one consonant with his aims and abilities, and from which the Institution would have derived a stimulus in its work, and lasting benefit. Mr Denny, going abroad early after his election, never had the opportunity, to which he must have looked forward so hopefully, of delivering his Presidential Address to a body whose objects are so completely kindred with what were his own individual aims. The present Session of the Institution was nearly closed at the time of Mr Denny's death at Buenos Ayres, on 17th March. An excellent and sympathetic Memoir, however, of the late President was at the request of the Council prepared and read by Mr John Ward, and which is now embodied, with a portrait of Mr Denny, in the present volume, in which the Members will find a detailed account of the numerous operations of a commercial, scientific, and philanthropic character in which their late President had from an early time been engaged.

Mr JOHN FERGUSON joined the Institution in 1865, at which time the Scottish Shipbuilders' Association, founded in 1860, became incorporated with the Institution. Mr Ferguson was a Member of Council of the Association during its First and Second Sessions, 1860-61 and 1861-62, and again in 1864-65. After the incorporation he was a Member of Council of the Institution during Sessions 1866-67 and 1867-68, and again in 1869-70, and in 1870-71 acted as Vice-President; in 1871-72 and 1872-73 he was a Member of Council. At the January Meeting in 1861 Mr Ferguson read a paper on "River Steamers" before the Scottish Shipbuilders' Association, in which he advocated the advantage of saloons to increase the comfort and extent of the cabin accommodation. The early advocacy, on Mr Ferguson's part, of this now common feature of our river steamers, largely helped the adoption of this improvement, and in April, 1865, he read a paper before the same Association on "The Comparative Merits of the Rules of Lloyds' and the Liverpool Underwriters' Association for the Construction of Iron Ships." Mr Ferguson took an active part in the negotiations carried on for the amalgamation of the Scottish Shipbuilders' Association and the Institution of Engineers, being a member of the Committee appointed in furtherance of this proposal. During Session 1865-66 of the now combined body Mr Ferguson read a paper on "The Safety and Seaworthiness of Vessels," the subject of which was carried into a discussion which followed on the Tonnage Laws, Mr Ferguson being appointed a member of Committee to approach the Board of Trade on the subject. In 1869 Mr Ferguson brought before the Institute a new form of Horizontal Propeller, the invention of Mr William Moodie, giving the results of experiments. Mr Ferguson's early training was that of a shipwright, and he was latterly managing partner of the shipbuilding works of Messrs Barclay, Curle & Co. whose works are now situated at Whiteinch.

Mr Ferguson took an active public interest in the neighbouring district, being Provost of the Burgh of Partick from 1875 to 1877, giving much attention to philanthropic movements, and as a member of the Parks Committee, assisted in the formation of the new Victoria Park in the Burgh. Mr Ferguson was born in Greenock in 1823, and died on 11th June, 1887.

Mr JAMES HUNTER joined the Institution in 1860. He was born at Muirkirk in 1818, and came to Coltness in 1836 to assist in drawing and carrying out plans for the construction of new Iron Works. His remarkable energy and ability were soon discovered, and in a very short time he became manager of the works. Two furnaces were built at first, but in 1846 they were increased to six, and ultimately reached twelve. Important and valuable seams of blackband ironstone were secured at Crofthead, some seven miles distant from Newmains, and the whole undertaking was managed with great technical skill and enterprise. Mr Hunter, however, found time to interest himself in all that pertained to the welfare of the community among whom he lived. The education of the young specially engaged his attention, and many years ago he devised a system by which the children of all his workmen should receive at the least possible expense the best possible elementary education. Both as Chairman of the School Board and in the management of the other Local Trusts, Mr Hunter showed much ability. For thirty years he was a Justice of the Peace in Lanarkshire, and at the time of his death he was a Deputy Lieutenant of Ayrshire. He was also a Director of the Scottish Widows' Fund and of the National Bank of Scotland. About a year ago he retired from the active management of Coltness Works. Mr Hunter died on 5th October, 1886.

Mr WILLIAM M'ONIE, Jun., was elected a Member of the Institution in 1883, and was a member of the firm of Messrs W. & W. M'Onie, Engineers, Glasgow. Mr M'Onie was born at Glasgow, and received his education in England, afterwards becoming apprenticed as an engineer to Messrs Randolph, Elder & Co., thereafter becoming a partner of the firm of Messrs W. & A. M'Onie, Engineers, Glasgow, and afterwards became principal in the new firm of Messrs W. & W. M'Onie, Engineers. Mr M'Onie took an active part in the inauguration of the scheme for the holding of an International Exhibition in Glasgow, one in which his position and abilities, together with the respect in which he was held, eminently qualified him. Mr M'Onie died at Glasgow, on 9th March, 1887, aged 36 years.

Mr JOHN THOMSON joined the Institution in 1876, and from the first took an active interest in the business of the meetings, holding office as a Member of Council during Sessions 1877-78 and 1878-79, he was one of the Vice-Presidents during Sessions 1879-80 and 1880-81, and was also Honorary Librarian during the latter Session.

In April, 1880, Mr Thomson read a paper on "The St. Petersburg Water Works," and for which he was awarded the Institution Medal. He also frequently took part in the discussion of papers read, and in 1883 represented the Institution at the Board of Management of the College of Science and Arts, and latterly in the same capacity as a Governor of the Glasgow and West of Scotland Technical College. Mr Thomson's early training as an engineer was obtained with Messrs A, & J. Inglis, and at the Hyde Park Works, Glasgow. He then extended his engineering experience by making several voyages to Brazil as engineer with the Royal Mail Steam Packet Co, afterwards becoming manager to the Duke of Bedford's water, gas, and sewage works at Thorney in the Fens. Latterly he became manager with Messrs R. Laidlaw & Son, Ironfounders, Glasgow, afterwards being assumed as a partner, and where his engineering skill and commercial ability were conspicuously shown in the many important contracts carried out by that firm. Mr Thomson took a lively interest in educational matters, especially those of a technical nature, and not only by his special abilities for such work, but by his genial and kindly manner, did much to stimulate others with whom he came in contact. Mr Thomson was born at Auchterarder, Perthshire, and died at Glasgow on 9th May, 1887, aged 50 years.

Mr ROBERT WYLIE was a native of Saltcoats, Ayrshire, where he received his early education. He became a Graduate of the Institution in 1873, and was elected a Member in 1884. He served his apprenticeship in the engineering department of Messrs Elder's Works, being afterwards assistant manager at Fairfield. Leaving Glasgow for Hartlepool, he became manager of the Hartlepool Engine Works, belonging to Messrs Thomas Richardson & Sons, where his ability and practical knowledge enabled him to carry out important works executed by that firm. Mr Wylie died 21st July, 1886, at the age of 36 years, from the result of an accident whilst engaged in his professional work.

Mr IVAN MAVOR became a Graduate of the Institution in 1880. He became an apprentice to Messrs John Elder & Co., Fairfield, in 1876, and afterwards was with Messrs A. & J. Inglis, at Pointhouse, thereafter being appointed draughtsman with Sir W. G. Armstrong, Mitchell & Co., Newcastle-on-Tyne, and shortly afterwards was appointed outside manager. In 1885 he became Consulting Naval Architect and General Superintendent of the Works of Messrs Hawthorn, Leslie, & Co., at Hebburn-on-Tyne, which position he held at his death. In 1885 his special knowledge of the question of the Stability of Ships was shown in his evidence given before the Wreck Commissioners. Mr Mavor showed great enthusiasm in his work, and during his residence in Glasgow, of which city he was a native, took an active part in the business of the Graduate Section of the Institution. Still quite a young man, he met his death at the age of 27, by accident, arising from the explosion of petroleum gas on board the S.S. "Petriana," at Birkenhead, on 26th December, 1886.

REPORT OF THE LIBRARY COMMITTEE.

IN accordance with the Bye-Laws, your Committee has much pleasure in reporting that Members, Associates, and Graduates have taken advantage of the Library of the Institution for reading and reference, and of the Library of the Philosophical Society for reference, to a much greater extent during the session now closing than in any previous one.

The annual scrutiny shows that 431 books have been lent out to 334 borrowers, but it is impossible to state the number of books which were used for reference. The books still out are coming in in a satisfactory manner, with the exception of 4 volumes, 2 pamphlets, and 1 part, which your Committee look upon as irrecoverable, having been out since 1873, notwithstanding repeated efforts to recover them.

During the past session 73 volumes, 29 parts, and 4 pamphlets have been added. Six of these volumes were presented and notified in the Monthly Transactions, but special mention might now be made of the Album of the Tay Viaduct presented by Mr William Arrol, the contractor, and of the Star Catalogue, an important book of reference, presented by Professor Piazzi Smyth, Astronomer· Royal for Scotland.

On behalf of the Institution the Committee now beg to tender its best thanks for the presentations made.

In accordance with the motion passed at the General Meeting held 26th October, 1886, books of recent publication (included in the 73 already mentioned) to the value of £20, have been added this session and detailed in the usual way in the Monthly Transactions.

The Institution exchanges Transactions with forty-one scientific societies. Fifteen weekly, two fortnightly, fourteen monthly, and four quarterly periodicals are received regularly in exchange for the

Transactions also. The greater number of the periodicals are regularly bound, and make 50 volumes, the total number of books and periodicals bound during the session being 37.

There are 1305 bound and 270 unbound volumes in the Library.

It may be mentioned that Members of all classes have the privilege of using, for reference, the Library of the Philosophical Society.

With the view of increasing the number of books in the Library and rendering it more useful, your Committee think it advisable that a suggestion be made to each member on his election to present to the Institution a drawing, plan, or model of engineering interest, or some scientific work for the Library; the names of all persons presenting any addition to the Library or to the collection of Plans and Models to be recorded and published as benefactors to the Institution.

CHAS. C. LINDSAY, *Librarian,*
Convener.

22nd March, 1887.

The following Form of Will is suggested for the purpose of bequeathing to the Institution any Books, Drawings, or Property which the donor may decide upon :—

"I give and bequeath to THE INSTITUTION OF ENGINEERS AND SHIPBUILDERS IN SCOTLAND (Incorporated),

(here specify the property bequeathed,)

and I hereby declare that the receipt of the Treasurer of the said Institution for the time being shall be a complete discharge to my executors for the said bequest."

DONATIONS TO LIBRARY.

Annual Report—British Association. 1885.

" Canadian Economics." From the Council of the British Association.

" Edinburgh Astronomical Observations," Vol. XV., Star Catalogue. From Professor Piazzi Smyth, F.R.S.E., Astronomer Royal for Scotland.

" A Text Book on Steam and Steam Engines," by Andrew Jamieson, Esq., C.E., Principal, College of Science and Arts, Glasgow. From the Author.

Album of Views of the Tay Bridge. From William Arrol, Esq.

Proceedings, Engineering Association of New South Wales. From the Association.

NEW BOOKS ADDED TO LIBRARY.

Riveted Joints, by Stoney.

Theory of Stresses, &c., by Stoney.

Busley—Die Schippmachine.

Bach—Die Wasseräder.

Kemp—Yacht Architecture.

History of the Invention of the Locomotive.

Sinclair—Locomotive Engine Management.

Matheson—Depreciation of Factories.

Gillespie on Roads.

Young—Strains in Girders and Roofs.

Shields on Girders and Roofs.

Seaton—Marine Engineering.

Life of James Nasmyth.

Barnaby—Marine Propellers.

Dictionary of Mechanics.

Smith—Coal and Coal Mining.
Anderson—Lightning Conductors.
Williams—Manual of Telegraphy.
Electricity as a Motive Power, Moncel and Geraldy.
The Works Manager's Handbook, Hutton.
Metalliferous Minerals, by D. C. Davies.
The Prospector's Handbook, by J. W. Anderson.
Moulder's and Founder's Guide, by T. Overman.
Aid to Survey Practice, by Jackson.
The Iron and Coal Trade Industries of the United Kingdom by
Meade.
Modern Practice of Shipbuilding in Iron and Steel, Thearle.

THE INSTITUTION EXCHANGES TRANSACTIONS WITH THE FOL-
LOWING SOCIETIES, &C. :—

Institution of Civil Engineers.
Institution of Civil Engineers of Ireland.
Institution of Mechanical Engineers.
Institution of Naval Architects.
Institute of Mining and Mechanical Engineers.
Iron and Steel Institute.
Liverpool Polytechnic Society.
Liverpool Engineering Society.
Literary and Philosophical Society of Manchester.
Midland Institute of Mining, Civil, and Mechanical Engineers
Mining Institute of Scotland.
Patent Office, London.
Philosophical Society of Glasgow.
Royal Scottish Society of Arts.
Royal Dublin Society.
South Wales Institute of Engineers.
Society of Engineers.
Society of Arts.

Manchester Association of Engineers.
North-East Coast Institution of Engineers and Shipbuilders.
The Sanitary Institute of Great Britain.
The Hull and District Institution of Engineers and Naval Architects.
American Society of Civil Engineers.
American Society of Mechanical Engineers.
Geological Survey of Canada.
The Canadian Institute, Toronto.
Master Car Builders' Association, U.S.A.
The Technical Society of the Pacific Coast, U.S.A.
Smithsonian Institution, U.S.A.
Bureau of Steam Engineering, Navy Department, U.S.A.
Royal Society of Tasmania.
Royal Society of Victoria.
Royal Academy of Sciences, Lisbon.
Société des Ingénieurs Civils de France.
Société Industrielle de Mulhouse.
Société d'Encouragement pour l'Industrie Nationale.
Société des Anciens Elèves des Ecôles Nationales d'Arts et Metiers.
Société des Sciences Physiques et Naturelles de Bordeaux.
Austrian Engineers' and Architects' Society, Vienna.
Engineers and Architects' Society of Naples.
The Association of Civil Engineers of Belgium.

COPIES OF THE TRANSACTIONS ARE FORWARDED TO THE
FOLLOWING COLLEGES, LIBRARIES, &C.:—

Glasgow University.	Stirling's Library, Glasgow.
University College, London.	Dumbarton Free Library.
M'Gill University, Montreal.	Lloyds' Office, London.
Stevens Institute of Technology, U.S.A.	Underwriters' Rooms, Glasgow.
	Do. Liverpool.
Cornell University, U.S.A.	Mercantile Marine Service Asso-
Mitchell Library, Glasgow.	ciation, Liverpool.

The Yorkshire College, Leeds.

PUBLICATIONS RECEIVED PERIODICALLY IN EXCHANGE FOR
INSTITUTION TRANSACTIONS :—

Annales Industrielles.	Journal de l'Ecole Polytechnic.
Colliery Guardian.	Nature.
Engineering.	Revue Industrielle.
Indian Engineering.	The Engineer.
Industries.	The Indian Engineer.
Iron.	The Machinery Market.
Iron and Coal Trades' Review.	The Marine Engineer.

The American Manufacturer and Iron World.

The Contract Journal. The Mechanical World. Stahl und Eisen.
Revue Maritime et Coloniale, Paris. L'Industria.

The Library of the Institution, at the Rooms, 207 Bath Street, is open daily from 9-30 a.m. till 8 p.m. ; on Meeting Nights of the Institution and Philosophical Society, till 10 p.m. ; and on Saturdays till 2 p.m. Books will be lent out on presentation of Membership Card to the Sub-Librarian.

Members have also the privilege of consulting the Books in the Library of the Philosophical Society.

The use of Library and Reading Room is open to Members, Associates, and Graduates.

The Library is open during Summer from 9-30 a.m. till 5 p.m.; and on Saturdays till 2 p.m.

The LIBRARY COMMITTEE are desirous of calling the attention of Readers to the " Recommendation Book," where entries can be made of titles of books suggested as suitable for addition to Library.

Copies of Library Catalogue, price 1s each, may be had at the Library, 207 Bath Street, or from the Secretary.

The Portrait Album lies in the Library for the reception of Members' Portraits.

Members are requested when forwarding Portraits to attach Signature to bottom of Carte.

The Council, being desirous of rendering the Transactions of the Institution as complete as possible, earnestly request the co-operation of Members in the preparing of Papers for reading and discussion at the General Meetings.

Early notice of such Papers should be sent to the Secretary, so that the dates of reading may be arranged.

Annual Subscriptions are due at the commencement of each Session, viz. :—

MEMBERS, £1 10s; ASSOCIATES, £1; GRADUATES, 10s.

LIFE MEMBERS, £20; LIFE ASSOCIATES, £15.

Membership Application Forms can be had at the Secretary's Office, 261 West George Street, or from the Sub-Librarian at the Rooms, 207 Bath Street.

Copies of the Reprint of Vol. VII., containing Paper on "The Loch Katrine Water Works," by Mr J. M. Gale, C.E., may be had from the Secretary. Price to Members, 7/6.

Members of this Institution, who may be temporarily resident in Edinburgh, will, on application to the Secretary of the Royal Scottish Society of Arts, at his Office, 117 George Street, be furnished with Billets for attending the Meetings of that Society.

The Meetings of the Royal Scottish Society of Arts are held on the 2nd and 4th Mondays of each Month, from November till April, with the exception of the 4th Monday of December.

LIST

OF

HONORARY MEMBERS, MEMBERS, ASSOCIATES, AND GRADUATES

OF THE

Institution of Engineers and Shipbuilders in Scotland

(INCORPORATED),

SESSION 1886-87.

HONORARY MEMBERS.

JAMES PRESCOTT JOULE, LL.D., F.R.S., 12 Wardle Road, Sale, near Manchester.

Professor CHARLES PIAZZI SMYTH, F.R.S.E., Astronomer-Royal for Scotland, 15 Royal Terrace, Edinburgh.

Professor Sir WILLIAM THOMSON, A.M., LL.D., D.C.L., F.R.SS.L. and E., Professor of Natural Philosophy in the University of Glasgow.

Professor R. CLAUSIUS, the University, Bonn, Prussia.

Sir JOSEPH WHITWORTH, Bart., C.E., LL.D., F.R.S., Manchester.

Professor JOHN TYNDALL, D.C.L., LL D., F.R.S., &c., Royal Institution, London.

HIS GRACE THE DUKE OF SUTHERLAND, Trentham, Stoke-upon-Trent.

Lord ARMSTRONG, C.B., LL.D., D.C.L., F.R.S., Newcastle on-Tyne.

Professor H. VON HELMHOLTZ, Berlin.

MEMBERS.

Names marked thus * were Members of Scottish Shipbuilders Association at Incorporation with Institution, 1865.

Names marked thus + are Life Members.

1881, Oct. 25: Allan W.   Baird,   Eastwood Villa, St. Andrew's Drive, Pollokshields.

1880, Feb. 24: William N. Bain,   Collingwood, Pollokshields. Glasgow.

1873, Apr. 22: H. W.   Ball,   Averley, Gt. Western Road, Hillhead, Glasgow.

1876, Jan. 25: James   Barr,   45 Oakshaw Street, Paisley.

1882, Mar. 21: Prof. Archd. Barr, B.Sc., C.E., The Yorkshire College, Leeds.

1868, Apr. 22: Edward   Barrow,   Rue de la Province, Sud, Antwerp, Belgium.

1881, Mar. 22: George H.   Baxter,   Ramage & Ferguson, Leith,

G. 1877, Nov. 20: M. 1887, Apr. 6: J. T.   Baxter,   9 Brighton Ter., Govan.

G. 1871, Feb. 21: M. 1865, Nov. 24: William S. Beck,   246 Bath Street, Glasgow.

1875, Jan. 26: Charles   Bell,   21 Victoria Place, Stirling.

David   *Bell,   Shipbuilder, Yoker, near Glasgow.

1880, Mar. 23: Imrie   Bell, C.E.,   1 Victoria Street, Westminster, London, S.W.

G. 1883, Mar. 20: M. 1884, Nov. 25: Andrew S. Biggart, C.E., Forth Bridge Works, South Queensferry.

1884, Mar. 25: John Harvard Biles,   Clydebank Shipyard, near Glasgow.

1866, Dec. 26: Edward   Blackmore,   Rookwood Road, Stamford Hill, London, N.

1864, Oct. 26: Thomas   Blackwood,   Shipbuilder, Port-Glasgow.

1869, Feb. 17: Geo. M'L.   Blair,   127 Trongate, Glasgow.

1867, Mar. 27: James M.   Blair,   2 Bute Gardens, Hillhead, Glasgow.

1883, Jan. 23: Chas. C.   Bone, C.E.,   23 Miller Street, Glasgow.

1888, Oct. 23: William L.   Bone,   Ant and Bee Works, West Gorton, Manchester.

1874, Jan. 27: Howard	Bowser,	13 Royal Crescent, W., Glasgow.
1880, Mar. 23: James	Brand, C.E.,	109 Bath Street, Glasgow.
G. 1873, Dec. 23: } M.1884, Jan. 22: } James	Broadfoot,	55 Finnieston St., Glasgow.
1865, Apr. 26: Walter	*Brock,	Engine Works, Dumbarton.
1859, Feb. 16: Andrew	*Brown,	London Works, Renfrew.
G. 1876, Jan. 25: } M. 1885, Nov.24: } Andrew M'N. Brown,		Castlehill House, Renfrew.
1886, Mar.23: George	Brown,	Comely Park, Dumbarton.
1885, Apr. 28: Walter	Brown,	Castlehill, Renfrew.
1880, Dec. 21: William	Brown,	Albion Works, Woodville Street, Govan, Glasgow.
G.1874, Jan. 27: } William M.1884, Jan. 22: }	Brown,	Houston Terrace, Paisley Road, Renfrew.
1858, Mar. 17: James	Brownlee,	23 Burnbank Gardens, Glasgow.
1860, Dec. 26: James C.	Bunten,	100 Cheapside St.,Glasgow.
1866, Apr. 26: Amedee	Buquet, C.E.,	15 Chemiss, St. Martin, Pontoise, S. O. France.
G. 1872, Oct. 22: } Hartvig M.1885, Nov.24: }	Burmeister,	Burmeister & Wain, Copenhagen, Denmark.
1880, Dec. 21: James W.	Burns,	5 Cecil Street, Paisley Rd., W., Glasgow.
1881, Mar.22: Thomas	Burt,	371 New City Rd., Glasgow.
1884, Jan. 22: Edward H.	Bushell,	G 19 Exchange Buildings, Liverpool.
1878, Oct. 29: Edward B.	Caird,	20 Lyndoch St., Glasgow.
1878, Dec. 17: James	Caldwell,	130 Elliot Street, Glasgow.
1885, Mar. 24: John B.	Cameron,	160 Hope Street, Glasgow.
1875, Dec. 21: J. C.	Cameron,	24 Pollok Street, Glasgow.
1886, Jan. 26: Andrew	Campbell,	53 Crookston Street, S.S., Glasgow.
1868, Dec 28: David	Carmichael,	Ward Foundry, Dundee.

1859, Nov. 23: Peter	Carmichael,	Dens Works, Dundee.
1862, Jan. 8: John	Carrick,	6 Park Quadrant, Glasgow.
1881, Nov. 22: John H.	Carruthers,	Craigmore, Queen Mary Avenue, C'shill, Glasgow.
1859, Oct. 26: Robert	Cassels,	168 St. Vincent Street, Glasgow.
1867, Jan. 30: Albert	Castel,	3 Lombard Court, London, E.C.
1883, Jan. 23: John	Clark,	British India Steam Navigation Co., 16 Strand, Calcutta.
1875, Oct. 26: W. J.	Clark,	Southwick, near Sunderland.
1880, Nov. 2: James	Clarkson.	Maryhill Engine Works, Maryhill, Glasgow.
1860, Apr. 11: James	Clinkskill,	1 Holland Place, Glasgow.
1884, Feb. 26: James T.	Cochran,	Duke Street, Birkenhead.
1881, Oct. 25: George	Cockburn,	Rhodora Villa, St. Andrew's Drive, Pollokshields, Glasgow.
G. 1876, Dec. 19: } M. 1884, Mar. 25: } Charles	Connell,	Whiteinch, Glasgow.
G. 1877, Dec. 18: } M. 1885, Nov. 24: } James	Conner,	Isle of Wight Railway, Sandown, England.
Original: Robert	Cook,	Woodbine Cottage, Pollokshields, Glasgow.
1864. Feb. 17: James	Copeland,	16 Pulteney St., Glasgow.
1864, Jan. 20: William R.	Copland, C.E.,	146 West Regent Street, Glasgow.
1868, Mar. 11: S. G. G.	Copestake,	Glasgow Locomotive Works, Little Govan, Glasgow.
1866, Nov. 28: M'Taggart	Cowan, C.E.,	109 Bath Street, Glasgow.
1868, Apr. 22: David	Cowan, C.E.,	Mount Gerald House, Falkirk.
1861, Dec. 11: William	Cowan,	46 Skene Terrace, Aberdeen.

1883, Dec. 18: Samuel	Crawford,	Clydebank, near Glasgow.
1881, Mar. 22: William	Crockatt,	2 Marjory Place, Pollok-shields, Glasgow.
1866, Dec. 26: James L.	Cunliff,	Plewlands House, Merchiston, Edinburgh.
1872, Nov. 26: David	Cunningham, C.E.,	Harbour Chambers, Dundee.
1884, Dec. 23: Peter N.	Cunningham,	5 North East Park Street, Glasgow.
1869, Jan. 20: James	Currie,	16 Bernard Street, Leith.
G. 1874, Feb. 24:⎫ James M. 1882, Dec. 19:⎬	Davie,	234 Cathcart Road, Crosshill, Glasgow.
1861, Dec. 11: Thomas	Davison,	248 Bath Street, Glasgow.
1864, Feb. 17: St. J. V.	Day, C.E.,	115 St. Vincent Street, Glasgow.
1869, Feb. 17: James	Deas, C.E.,	Engineer, Clyde Trust, 7 Crown Gardens, Glasgow.
1882, Dec. 19: J. H. L. Van Deinse,		85 de Ruyterkade, Amsterdam.
1883, Nov. 21: James	Denholm,	360 Dumbarton Road, Glasgow.
1866, Feb. 14: A. C. H.	Dekke,	Shipbuilder, Bergen, Norway.
Peter	*Denny,	Helenslee, Dumbarton.
1873, Feb. 18: William (*President.*)	Denny,	Leven Shipy'd, Dumbarton.
G. 1873, Dec. 23:⎫ Peter M. 1884, Jan. 22:⎬	Dewar,	163 Sandyford St., Glasgow.
1878, Mar. 19:Frank W.	Dick,	Hallside Steel Works, Newton.
G. 1873, Dec.24:⎫ James S. M.1878, Jan. 22:⎬	Dixon,	170 Hope Street, Glasgow.
1882, Nov. 28: John G.	Dobbie,	British India Steam Navigation Co., Mazagon Dockyard, Bombay.

1871, Jan. 17: William   Dobson,   The Chesters, Jesmond, Newcastle-on-Tyne.

1864, Jan. 20: James   Donald,   Abbey Works, Paisley.

1876, Jan. 25: James   Donaldson,   Almond Villa, Renfrew.

1863, Nov.25: Robert   Douglas,   Dunnikier Foundry, Kirkcaldy.

1886, Nov. 23: Patrick   Doyle, C.E., 19 Lall. Bazar, Calcutta.

1884, Dec. 23: John W.W. Drysdale,   5 Whitehill Gardens,G'gow.

1882, Oct. 24: Chas. R.   Dubs,   Glasgow Locomotive Works, Glasgow.

1886, Nov. 23: John   Duncan,   Ardenclutha,Port-Glasgow.

1887, Apr. 6: D. J. Russell Duncan, C.E., 10 Airlie Gardens, Kensington, London, W.

1864, Oct 26: Robert   *Duncan,   Shipbuilder, Port-Glasgow.
(Past President.)

1881, Jan. 25: Robert   Duncan,   Whitefield Engine Works, Govan, Glasgow.

1873, Apr. 22: Robert   Dundas,C.E., 3 Germiston Street,Glasgow.
(Vice President.)

1869, Nov. 23: David Jno. Dunlop,   Inch Works, Port-Glasgow.

1877, Jan. 23: John G.   Dunlop,   17 Goulton Road, Lower Clapton, London.

1880, Mar. 23: Hugh S.   Dunn,   Earlston Villa, Caprington, Kilmarnock.

1886, Oct. 26: Peter   Dunn, C.E., Endrick Villa,Pollokshields, Glasgow.

1883, Oct. 23: Henry   Dyer, C.E., MA., 8 Highburgh Terrace,
(Member of Council.)   Dowanhill, Glasgow.

1876, Oct. 24: Jn. Marshall Easton,   Redholm, Helensburgh.

1885, Feb. 24: Francis   Elgar, LL.D., Director, H. M. Dockyard,
(Member of Council.)   Admiralty, London, S.W.

1875, Oct. 26: James G.   Fairweather,C.E.,B.Sc., 21 St. Andrew Square, Edinburgh.

John	*Ferguson,	Shipbuilder, Whiteinch, Glasgow.
G. 1869, Nov. 23: M.1878, Mar. 19: } John	Ferguson,	Shipbuilder, Leith.
1874, Feb. 24: Immer	Fielden,	2 Thornton villas, Holderness Road, Hull.
1880, Jan. 27: Alexander	Findlay,	Hamilton Road, Motherwell.
G. 1873, Dec. 23: M.1884, Nov. 25: } E. Walton	Findlay,	Ardeer, Stevenston.
1884, Dec. 23: Finlay	Finlayson,	Alexandria Place Colt Terrace, Coatbridge.
Original: William	Forrest,	66 Bath Street, Glasgow.
1872, Nov. 26: Thomas	Forrest, M.E.,	Dumfries Ironworks, Dumfries.
1883, Dec. 18: Lawson	Forsyth,	10 Grafton Sq., Glasgow.
1870, Jan. 18: William (*Member of Council.*)	Foulis,	Engineer, Corporation Gas Works, 42 Virginia St., Glasgow.
1880, Nov. 2: Samson	Fox,	Leeds Forge, Leeds.
1879, Nov. 25: John	Frazer,	P. Henderson & Co., 15 St. Vincent Place, Glasgow.
1885, Jan. 27: Peter	Fyfe,	1 Montrose St., Glasgow.
1858, Nov. 24: James M. (*Past President; Member of Council, and Treasurer.*)	Gale, C.E.,	Engineer, Corporation Water Works, 23 Miller Street, Glasgow.
1862, Jan. 8: Andrew	Galloway, C.E.,	St. Enoch Station, Glasgow.
1883, Oct. 23: Gilbert H.	Garrett,	Robert Stephenson & Co., South Street, Newcastle-on-Tyne.
1873, Dec. 23: Bernard	Gatow,	Veritas Office, 29 Waterloo Street, Glasgow.
G. 1873, Dec. 23: M.1882, Mar.21: } Andrew	Gibb,	Rait & Gardiner, Millwall Docks, London.
1886, Nov. 23: Paterson	Gifford,	6 Union Street, Glasgow.

1859, Nov. 23: Archibald	*Gilchrist,	11 Sandyford Pl., Glasgow.
G. 1866, Dec. 26:) James M.1878, Oct. 29:)	Gilchrist,	Stobcross Engine Works, FinniestonQuay,Glasgow.
1859, Dec. 21: David C.	Glen,	14 Annfield Place, Dennis- toun, Glasgow.
1868, Nov. 25: Thomas	Goldie,	Waverley Mills, Ceres Road, Cape of Good Hope.
1864, Feb. 17: James	Goodwin.	Ironfounder, Ardrossan.
1866, Mar. 28: Gilbert S.	Goodwin,	Alexandra Buildings,James Street, Liverpool.
1868, Mar. 11: Joseph	Goodfellow,	136 Sackville Place, Stir- ling Road, Glasgow.
1882, Apr. 25: H. Garrett	Gourlay,	Dundee Foundry, Dundee.
Edwin	*Graham,	Osbourne, Graham, & Co., Hylton, Sunderland.
1858, Mar. 12: George	Graham,C.E.,	Engineer, Caledonian Rail- way, Glasgow.
1876, Jan. 25: Thomas M.	Grant,	4 Clayton Terrace, Dennis- toun, Glasgow.
1871, Mar. 28: Thomas	Gray,	Chapel Colliery, Newmains.
1862, Jan. 8: James	Gray,	Pathhead Colliery, Cum- nock, Ayrshire.
1870, Feb. 22: P. B. W.	Gross, M.E.,	4 Albion Place, Cumberland Road, Bristol.
1881, Dec. 20: L. John	Groves,	Engineer, Crinan Canal, Ar- drishaig.
1879, Nov. 25: Robert	†Hadfield,	Hadfield Steel Foundry Co., Attercliffe, Sheffield.
1872, Feb. 27: A. A.	Haddin, C.E.,	131 West Regent Street, Glasgow.
1881, Jan. 25: William	Hall, jun.,	Shipbuilder, Aberdeen.
1876, Oct. 24: David	Halley,	Burmeister & Wain, Copen- hagen, Denmark.
G. 1874, Feb. 24:) M.1885, Nov.24:) Archibald	Hamilton,	New Dock Works, Govan.

G. 1873, Dec. 23:} David C. M.1881, Nov. 22:}	Hamilton,	Clyde Shipping Co., 21 Carlton Place, Glasgow.
G. 1866, Dec. 26:} James M.1873, Mar. 18:}	Hamilton,	Jun., Ardedynn, Kelvinside, Glasgow.
John	*Hamilton,	22 Athole Gardens, Gl'gow.
G. 1869, Nov. 23:} J. B. M.1875, Feb. 23.}	Hamond,	Didsbury, Manchester.
1876, Feb. 22: Walter	Hannah,	Barwood, Helensburgh.
G. 1880, Nov. 2:} Bruce M.1884, Jan. 22:}	Harman,	Lancefield House, Lancefield Street, Glasgow.
1878, Mar. 19: Timothy	Harrington,	61 Gracechurch Street, London, E.C.
1875, Jan. 26: Peter T.	Harris,	19 West St. (S.S.),Glasgow.
G. 1874, Feb. 24:} C. R. M.1880, Nov. 23.}	Harvey,	166 Renfrew St., Glasgow.
1887, Feb. 22: John H.	Harvey,	Benclutha, Port-Glasgow.
1864, Nov. 23: John	Hastie,	Kilblain Engine Works, Greenock.
1871, Jan. 17: William	Hastie,	Kilblain Engine Works, Greenock.
1879, Nov. 25: A. P.	†Henderson,	30 Lancefield Quay, Glasgow.
1877, Feb. 20: David	*Henderson,	Meadowside, Partick, Glasgow.
1873, Jan. 21: John	†Henderson,	Jun., Meadowside, Partick, Glasgow.
1879, Nov. 25: John L.	†Henderson,	Westbank House, Partick, Glasgow.
1878, Dec. 17: William	Henderson,	Meadowside, Partick, Glasgow.
1880, Nov. 2: William	Henderson,	C.E., 121 West Regent Street, Glasgow.
1870, May 31: Richard	Henigan,	C.E., Alma Road, Cotford Place, Southampton.
1877, Feb. 20: George	Herriot,	7 York Street, Glasgow

	Laurence	*Hill, C.E.,	5 Doon Gardens, Hillhead, Glasgow.
1886, Mar. 23:	Geo. Lisle	Hindmarsh,	Lloyds Surveyor, Cardiff.
1880, Nov. 2:	Charles P.	Hogg, C.E,	175 Hope Street, Glasgow.
	(*Vice President.*)		
1883, Mar. 20:	John	Hogg,	Victoria Engine Works, Airdrie.
1880, Mar. 23:	F. G.	Holmes, C.E.,	103 Bath Street, Glasgow.
1883, Mar. 20:	Matthew	Holmes,	8 Annfield Terrace, West Partick
Original:	James	Howden.	8 Scotland Street, Glasgow.
1884, Apr. 22:	John G.	Hudson,	18 Aytoun Road., Pollokshields, Glasgow.
Original:	Edmund	*Hunt,	87 St. Vincent St., Glasgow.
1881, Jan. 25:	James	Hunter,	Aberdeen Iron Works, Aberdeen.
G. 1873, Dec. 23: M. 1885, Nov. 24:	Guybon	Hutson,	Kelvinhaugh Engine Works, Glasgow.
G. 1873, Dec. 23: M. 1877, Feb. 20:	P. S.	Hyslop,	14 Leven Street, Pollokshields.
Original:	John	*Inglis,	64 Warroch St., Glasgow.
1861, May 1:	John	Inglis, Jun.,	Point House Shipyard, Glasgow.
	(*Vice President.*)		
1879, Jan. 21:	Thos. F.	Irwin,	2A Tower Chambers, Old Churchyard, Liverpool.
1880, Nov. 2:	Lawrence N.	Jackson,	C/o The Manager Galle Face Hotel, Colombo, Ceylon.
1875, Dec. 21:	William	Jackson,	Govan Engine Works, Govan, Glasgow.
1884, Jan. 22:	J. Yate	Johnson, C.E.,	115 St. Vincent Street, Glasgow.
1879, Feb. 25:	David	Johnston,	6 Osborne Place, Copeland Road, Govan, Glasgow.
1870, Dec. 20:	David	Jones,	Highland Rlwy., Inverness.

1883, Jan. 23: F. C.   Kelson,   Angra Bank, Waterloo Park, Waterloo, Liverpool.

1872, Mar. 26: Ebenezer   Kemp,   Linthouse Engine Works, Govan, Glasgow.

1875, Nov. 23: William   Kemp,   EllenSt. Engineering Works, Govan, Glasgow.

1878, Mar. 19: Hugh   Kennedy.   Redclyffe, Partickhill, Glasgow.

1877, Jan. 23: John   Kennedy,   R. M'Andrew & Co., Suffolk House, Laurence Pountney Hill, London, E.C.

1876, Feb. 22: Thomas   Kennedy,   Water Meter Works, Kilmarnock.

1876, Oct. 24: Andrew   Kerr, C.E.,   Town Surveyor's Office, Warrnambool, Victoria, Australia.

           David   *Kinghorn,   172 Lancefield St., Glasgow.

1879, Dec. 23: John G.   Kinghorn,   Tower Buildings, Water Street, Liverpool.

1885, Nov. 24: Frank E.   †Kirby,   Detroit, U.S., America.

1864, Oct. 26: Alex. C.   Kirk,   19 Athole Gardens, Hillhead, Glasgow.
          (*Member of Council.*)

Original: David   *Kirkaldy,   Testing and Experimenting Works, 99 Southwark Street, London, S.E.

1885, Jan. 27: Charles A.   Knight,   107 Hope Street, Glasgow.

1880, Mar. 23: Frederick   Krebs,   Copenhagen, Denmark.

1875, Oct. 26: William   Laing,   17 M'Alpine St., Glasgow.

1858, Apr. 14: David   Laidlaw,   Chaseley, Skelmorlie, by Glasgow.

1884, Mar. 25: John   Laidlaw,   98 Dundas Street, S.S., Glasgow.

1862, Nov. 26: Robert   Laidlaw,   147 E. Milton St., Glasgow.

1880, Feb. 24: James   Lang,   ^c/_o George Smith & Sons, City Line, 45 West Nile Street, Glasgow.

1884, Feb. 26: John   Lang, Jun.,   Church Street, Johnstone.

Original: James G. *Lawrie,   2 Westbourne Terrace, *(Past President.)*   Glasgow.

G. 1873, Dec. 23: ⎱ Charles C. Lindsay,C.E.,167 St.VincentSt.,Glasgow.
M. 1876, Oct. 24: ⎰   *(Member of Council.)*

1884, Feb. 26: John   List,   D. Currie & Co., Blackwall, London, E.

1862, Apr. 2: H. C.   Lobnitz,   Renfrew.

1865, Dec. 20: John L.   Lumsden,   ————

1885, Oct. 27: John   Lyall,   69 St. Vincent Crescent, Glasgow.

1873, Jan. 21: James M.   Lyon, M.E.,   Engineer and Contractor, Singapore.

1884, Dec. 23: John   M'Beth,   5 Park Street, S.S., Glasgow.

1858, Feb. 17: David   M'Call, C.E.,   160 Hope Street, Glasgow.

1874, Mar. 24: Hector   MacColl,   19 St. Edmond's Road, Bootle, Liverpool.

   Hugh   *MacColl,   Manager, Wear Dock Yard, Sunderland.

1883, Oct. 23: James   M'Creath,C.E., 95 Bath Street, Glasgow.

1871, Jan. 17: David   M'Culloch,   Vulcan Works, Kilmarnock.

1884, Feb. 26: James   M'Ewan,   Cyclops Foundry, 50 Peel Street, London Road, Glasgow.

1880, Nov. 2: James W.   Macfarlane,   Valeview House, Overlee, Busby.

1886, Oct. 26: Walter   Macfarlane,   12 Lynedoch Cres., Glasgow.

G. 1874, Feb. 24: ⎱ George   M'Farlane,   65 Great Clyde Street,
M. 1885, Nov. 24: ⎰   Glasgow.

1886, Jan. 26: Thomas　　M'Gregor,　　10 Mosesfield Terrace, Springburn.

1887, Apr. 6: Edward　　Mackay,　　8 George Square, Greenock·

1881, Mar. 22: William A. Mackie,　　3 Broomhill Terrace, Partick, Glasgow.

1873, Jan. 21: J. B. Affleck M'Kinnel,　　Dumfries Iron Works, Dumfries.

1859, Dec. 21: Robert　　M'Laren,　　22 Canal St., S.S., Glasgow.

G. 1880, Nov. 2:　}
M. 1885, Dec. 22:　} Robert　M'Laren, Jun.,Eglinton Foundry,Glasgow.

　　　　　Sir Andrew *Maclean,　　Viewfield House, Partick, Glasgow.

G. 1874, Feb. 24:　}
M. 1885, Nov. 24:　} Andrew　M'Lean, Jun , Viewfield House, Partick.

1884, Dec. 23: James　　M'Lellau,　　10 W. Garden Street, Glasgow.

1858, Nov. 24: Walter　　M'Lellan.　　127 Trongate, Glasgow.

1886, Dec. 21: William T. †Maclellan,　　Clutha Iron Works, Glasgow.

　　　　　John　　*M'Millan,　　Shipbuilder, Dumbarton.

　　　　　William　　*MacMillan,　　19 Elgin Terrace, Partick. Glasgow.

1884, Dec. 23: John　　M'Neil,　　Helen St., Govan, Glasgow.

1883, Jan. 23: William　　M'Onie, Jun., 128 West Street, Glasgow.

1883, Jan. 23: James　　M'Ritchie, C.E , Singapore.

1864, Oct. 26: Robert　　*Mansel,　　Shipbuilder, Whiteinch,
　　　　(*Past President.*)　　Glasgow.

1875, Dec. 21: George　　Mathewson,　　Bothwell Works, Dunfermline.

1884, Apr. 22: Henry A.　Mavor,　　140 Douglas St., Glasgow.

1876, Jan. 25: William W. May,　　142 Fountain Road, Walton, Liverpool.

1887, Jan. 25: Henry　　Mechan,　　17 Fitzroy Place, Glasgow.

1883, Feb. 20: James　　Meek,　　————

G. 1876, Oct. 24.　}
M. 1882, Nov. 28.　} James　Meldrum,C.E., 3 Elmbank Street, Glasgow.

1883, Jan. 23: William	Melville,C.E.,	Caledonian Ry., Buchanan Street, Glasgow.
1881, Mar. 22: William	Menzies,	7 Dean Street, Newcastle-on-Tyne.
1861, Dec. 11: Daniel	Miller, C.E.,	204 St. Vincent St., Glasgow.
G. 1873, Dec. 23: ⎱ John F. M. 1881, Nov. 22: ⎰	Miller,	Greenoakhill, Broomhouse.
Original: James B.	Mirrlees,	45 Scotland St., Glasgow.
1886, Jan. 26: Alexander	Mitchell,	4 Bellevue Terrace, Spring-burn.
1876, Mar. 21: James	Mollison,	Lloyd's Register, 36 Oswald Street, Glasgow.
1869, Dec. 21: John	Montgomerie,	210 Great Northern Ter., Possil Park.
1883, Nov. 21: Joseph	Moore,	East Finchley, London.
1862, Nov. 26: Ralph	Moore, C.E.,	13 Clairmont Gardens, Glasgow.
1878, Apr. 23: Robert H.	Moore,	Mount Blue Works, Cam-lachie, Glasgow.
G. 1878, Dec. 17: ⎱ Robert M. 1883, Jan. 23 ⎰	Morton,	53 Waterloo St., Glasgow.
1885, Mar. 24: Edmund	Mott,	Board of Trade Surveyor, 7 York Street, Glasgow.
1864, Feb. 17: Hugh	Muir,	7 Kelvingrove Ter., Glasgow.
1882, Jan. 24: John G.	†Muir,	c/o 169 West George Street, Glasgow.
1870, Mar. 22: Wm. T.	Mumford,	36 Oswald Street, Glasgow.
1882, Feb. 21: George	Munro,	254 Bath Street, Glasgow.
1882, Dec. 19: Robert D.	Munro,	141 Buchanan St., Glasgow.
Original: James	Murdoch,	Shipbuilder, Port-Glasgow.
1880, Jan. 27: William	Murdoch,	20 Carlton Place, Glasgow.
1886, Jan. 26: James	Murray,	8 Brown Street, Glasgow.
1877, Jan. 23: Robert	Murray,	25A Coltman Street, Hull.
1881, Jan. 25: Henry M.	Napier,	Shipbuilder, Yoker, near Glasgow.

1857, Dec. 23: John   †*Napier.   23 Portman Sq., London.

1881, Dec. 20: Robert T. †Napier,   Shipbuilder, Yoker, near
                                                        Glasgow.

Original: Walter M. Neilson,   Queen's Hill Kirkcudbright-
                (Past President.)   shire.

1869, Nov. 23: Theod. L.   Neish,   ————

A. 1865, Apr. 26: } R. S.   *Newall, F.R.S., F.R.A.S., &c., Ferndene,
M. 1879, Oct. 28: }   Gateshead on-Tyne.

1883, Dec. 18: Thomas   Nicol,   Clydebank, near Glasgow.

1884, Dec. 23: Wm. H.   Nisbet,   Mavisbank, Partickhill,
                                                        Glasgow.

1887, Apr. 6: William   Nish,   15 Govan Road, Glasgow.

1876, Dec. 19: Richard   Niven, C.E.,   Dalnottar House, Old Kil-
                                                        patrick.

1861, Dec. 11: John   Norman,   475 New Keppochhill Road,
                                                        Glasgow.

1886, Jan. 26: George   Oldfield,   Greenlea, Johnstone.

1882, Jan. 24: Robert S.   Oliver, C.E., Highland   Railway   Co.,
                                                        Inverness.

1860, Nov. 28: John W.   Ormiston,   Shotts Iron Works, Shotts.

1885, Mar. 24: Alex. T.   Orr,   Hall, Russell, & Co., Aber-
                                                        deen.

1867, Apr. 24: T. R.   Oswald,   The Southampton Ship
                                                        building & Engineering
                                                        Works. Southampton.

1882, Mar. 21: Geo. S.   Packer, F.I.C., Hallside Steel Works,
                                                        Newton, near Glasgow

1888, Nov, 21: W. L. C.   Paterson,   19 St. Vincent Crescent,
                                                        Glasgow.

1877, Apr. 24: Andrew   Paul,   Levenford Works, Dum-
                                                        barton.

G. 1884, Feb. 26: } Matthew Paul, Jun.,   Levenford Works, Dum-
M. 1886, Dec. 21: }   barton.

1880, Nov. 2: James M.   Pearson, C.E., Strand Street, Kilmarnock.

1866, Dec. 26: Sir William Pearce, Bart., M.P., Fairfield Shipyard, Govan, Glasgow.

1868, Dec. 23: Eugène   Perignon,C.E., 105 Rue Faubourg, St. Honoré, Paris.

1887, Apr. 6: Theodor J. Poretchkin, Russian Imperial Navy, °/₀ Blackley Young & Co., 103 Holm St., Glasgow.

          John   *Price,   6 Osborne Villas, Jesmond, Newcastle-on-Tyne.

1877, Nov. 20: F. P.   Purvis,   Craig Villa, Dumbarton.

1868, Dec. 23: Henry M. Rait,   155 Fenchurch St.,London.

1873, Apr. 22: Richard Ramage,   Shipbuilder, Leith.

1872, Oct. 22: David   Rankine,   75 West Nile St., Glasgow.

1886, Mar. 23: John F. Rankin,   Eagle Foundry, Greenock.

1881, Jan. 25: Charles Reid,   Lilymount, Kilmarnock.

1883, Nov. 21: George W. Reid,   Highland Railway, Inverness.

1868, Mar. 11: James   Reid,   LocomotiveWorks, Springburn, Glasgow.
          (*Past President.*)

1869, Mar. 17: James   Reid,   Shipbuilder, Port Glasgow.

          John   *Reid,   Shipbuilder, Port-Glasgow.

1880, Apr. 27: John   Rennie,   Ardrossan Shipbuilding Co., Ardrossan.

G. 1873, Dec. 23:} Charles H. Reynolds,   Cuprum House, Hamilton Ter., Partick, Glasgow.
M. 1881, Nov. 22:}

1876, Oct. 24: Duncan   Robertson,   8 Brighton Place, Govan, Glasgow

1886, Apr. 27: James   Riley,   Steel Company of Scotland, 23 Royal Exchange Sq., Glasgow.

Original: James   Robertson,   95 Colmore Row, Birmingham.

1873, Jan. 21: John   Robertson,   Grange Knowe, Pollokshields, Glasgow.

1863, Nov. 25: William   Robertson, C.E., 123 St. Vincent Street, Glasgow.

1884, Apr. 22: R. A.   Robertson, 42 Aytoun Road, Pollokshields, Glasgow.

Original: Hazltn. R. *Robson, 14 Royal Cresct., Glasgow.
(*Past President.*)

1877, Feb. 20: Jno. MacDonald Ross, 11 Queen's Cres., Glasgow.

1861, Dec. 11: Richard G. Ross, 21 Greenhead St., Glasgow.

G. 1864, Nov. 23:⎱ Alex.   Ross, C.E., Lyunwood, Alva.
M. 1870, Jan. 18:⎰

Original: David   *Rowan, 231 Elliot Street, Glasgow.
(*Past President.*)

G. 1875, Dec. 21:⎱ James   Rowan, 231 Elliot Street, Glasgow.
M. 1885, Jan. 27:⎰

1877, Oct. 30: Alexander   Russell, 186 North Street, Glasgow.

G. 1858, Dec. 22:⎱ George   Russell, Engineer, Motherwell.
M. 1863, Mar. 4:⎰ (*Member of Council.*)

1881, Feb. 22: Joseph   Russell, Shipbuilder, Port-Glasgow.

1876, Oct. 24: Peter   Samson, Board of Trade Offices, Downing Street, London, S.W.

1885, Feb. 24: James   Samuel, Jun., 238 Berkeley St., Glasgow.

1883, Feb. 20: John   Sanderson, Lloyd's Registry, 36 Oswald Street, Glasgow.

1882, Dec. 19: Prof. Jas.   Scorgie, F.C.S., Civil Engineering College Poona, India.

1884, Apr. 22: Andrew   Scott, 56 Græme Street, Glasgow.

1872, Jan. 30: James E.   Scott, 13 Rood Lane, London.

1881, Jan. 25: John   Scott, Whitebank Engine Works, Kirkcaldy.

1860, Nov. 28: Thos. B.   *Seath, 42 Broomielaw, Glasgow.

1875, Jan. 26: Alexander   Shanks, Belgrade, Aytoun Road, Pollokshields, Glasgow.

1858, Nov. 24: William   Simons, Renfrew.

1862, Jan. 22: Alexander   Simpson, C.E., 175 Hope Street, Glasgow.
(*Member of Council.*)

1887, Jan. 25: Robert	Simpson, B.Sc.,	175 Hope St., Glasgow.
1871, Mar. 28: Hugh	Smellie,	Belmont Grange Terrace, Kilmarnock.
Original: Alexander	Smith,	57 Cook Street, Glasgow.
1880, Nov. 2: Alexr. D.	Smith,	85 Maxwell Road, Pollok-shields, Glasgow.
1869, Mar. 17: David S.	Smith,	Hellenic Steam Navigation Co., Syra, Greece.
G. 1868, Dec. 23:) Hugh M. 1874, Oct. 27:)	Smith,	Argyle Engine Works, Mansion St., Glasgow.
1870, Feb. 22: Edward	Snowball,	Engineer, Hyde Park Loco-motive Works, Spring-burn, Glasgow.
1887, Jan. 25: Peter A.	Somervail,	35 Burnbank Gardens, Glasgow.
1883, Oct. 23: Andrew	Sproul,	2 Springkell St., Greenock.
1886, Feb. 23: George	Stanbury,	81 Brisbane St., Greenock.
1883, Dec. 18: Alex. E.	†Stephen,	12 Park Terrace, Glasgow.
John	†*Stephen,	Linthouse, Govan, Glasgow.
1881, Nov. 22: Alex.	Steven,	Provanside, Glasgow.
1867, Jan. 30: Duncan	Stewart,	47 Summer Street, Glasgow.
1874, Oct. 27: Peter	Stewart,	53 Renfield Street, Glasgow.
G. 1873, Dec. 23:) W. B. M. 1882, Oct. 24:)	Stewart,	18 Newton Place, Glasgow.
1866, Nov. 28: James	Stirling,	Loco. Engineer, S. Eastern Ry., Ashford, Kent.
Original: Patrick	Stirling,	The Great Northern Rail-way, Doncaster.
1881, Jan. 25: Walter	Stoddart,	Caledonian Railway, Car-stairs.
1877, Jan. 23: James	Syme,	8 Glenavon Ter., Partick, Glasgow.
1879, Oct. 28: James	Tait, C.E.,	Wishaw.
1882, Apr. 25: Alex. M.	Taylor,	Java Cottage, Lenzie.

1885, Apr. 28: Peter   Taylor,   62 Queen Street, Renfrew.

1879, Mar. 25: Staveley   Taylor,   Russell & Co., Shipbuilders, Greenock.

1873, Dec. 23: E. L.   Tessier,   Veritas Office, 29 Waterloo Street, Glasgow.

1885, Jan. 27: George W. Thode,   107 Hope Street, Glasgow.

1887, Apr. 26: Arthur W. Thomson, B.Sc., 79 West Regent Street, Glasgow.

1882, Apr. 25: Geo. P.   Thomson,   Clydebank, Dumbartonshire.

1883, Dec. 18: George   Thomson,   9 Buckingham Ter., Partick, Glasgow.

1886, Mar. 23: James   Thomson, Jun., Leven Shipyard, Dumbarton.

1874, Nov. 24: Prof. James Thomson, C.E., LL.D., F.R.SS.L. & E., *(Past President)*   2 Florentine Gardens, Hillhead Street, Glasgow.

1868, Feb. 12: James M.   Thomson,   36 Finnieston St., Glasgow.

1882, Mar. 21: James R.   Thomson,   Clydebank, Dumbartonshire,

1868, May 20: John   Thomson,   36 Finnieston St., Glasgow.

1876, Feb. 22: John   Thomson,   147 East Milton Street, *(Member of Council.)*   Glasgow.

1875, Jan. 26: Robert S.   Thomson,   3 Melrose Street, Queen's Crescent, Glasgow.

1864, Feb. 17: W. R. M.   Thomson,   96 Buchanan St., Glasgow. gow.

1878, May 14: W. B.   Thompson,   Ellengowan, Dundee.

Original: Thomas C. Thorburn,   35 Hamilton Square, Birkenhead.

1874, Oct. 27: Prof. R. H. Thurston, M.E., C.E., Sibley College, Cornell University, Ithaca, N.Y., U.S.A.

1875, Nov. 23: John   Turnbull, Jun., Consulting Engineer, 255 Bath Street, Glasgow.

1876, Nov. 21: Alexander   Turnbull,   15 Whitehill Terrace, Dennistoun, Glasgow.

1880, Apr. 27: John	Tweedy,	Neptune Works, Newcastle-on-Tyne.
1865, Apr. 26: W. W.	Urquhart,	Blackness Foundry, Dundee.
1883, Jan. 23: Peter	Wallace,	Shipyard, Troon.
1885, Mar. 24: W. Carlile	Wallace,	Greenfield House, Dumbarton.
1886, Jan. 26: John	Ward,	Leven Shipyard, Dumbarton.
1875, Mar. 23: G. L. (*Member of Council.*)	Watson.	108 W. Regent St., Glasgow.
1864, Mar. 16: W. Renny (*Member of Council.*)	Watson,	16 Woodlands Ter., Glasgow.
1883, Jan. 23: D. W.	Watt,	19 Renfield Street, Glasgow.
. 1875, Dec. 21: } [.1886, Oct. 26: } R. G.	Webb,	c/o Fleming & Co., Bombay.
John	*Weild,	Underwriter, Exchange, Glasgow.
1874, Dec. 22: George	Weir, M.E.,	18 Millbrae Cres., Langside, Glasgow.
1874, Dec. 22: James	Weir, M.E.,	Silver Bank, Cambuslang. near Glasgow.
. 1876, Dec. 19: } [.1884, Feb. 26: } Thomas D.	Weir, C.E.,	97 W. Regent St., Glasgow.
1869, Feb. 17: Thomas M.	Welsh,	63 St. Vincent Cres., Glasgow.
1868, Dec. 23: Henry H.	West,	14 Castle Street, Liverpool.
1883, Feb. 20: Richard S.	White,	Shipbuilder, Armstrong, Mitchell, & Co., Newcastle on-Tyne.
1887, Apr. 6: James	Whitehead,	71 Scott Street, Glasgow.
1884, Nov. 25: John	Wildridge,	Consulting Engineer, Sydney, N.S.W., Australia.
1876, Oct. 24: Francis W.	Willcox,	45 West Sunniside, Sunderland.

1884, Dec. 23: James	Williamson,	Barclay, Curle, & Co., Whiteinch.
1883, Feb. 20: Robert	Williamson,	Lang & Williamson, Engineers, &c., Newport, Mon.
1878, Oct. 29: Thomas	Williamson,	————
Alex. H.	*Wilson,	Aberdeen Iron Works, Aberdeen.
1868, Dec. 23: James	Wilson, C.E.,	Water Works, Greenock.
1870, Feb. 22: John	Wilson,	165 Onslow Drive, Dennistoun, Glasgow.
1858, Jan. 20: Thomas	†*Wingate,	Viewfield, Partick.
G. 1873, Dec. 23: ⎫ Robert M. 1884, Jan. 22: ⎭	Wyllie,	Hartlepool Engine Works, Hartlepool.
1867, Nov. 27: John	Young.	Galbraith Street, Stobcross, Glasgow.

ASSOCIATES.

Thomas	*Aitken,	8 Commercial Street, Leith.
1883, Oct. 28: John	Barr,	Secretary to Glenfield Co., Kilmarnock.
1882, Dec. 19: Wm.	Begg,	Gartfern, High Crosshill, Rutherglen.
1884, Dec. 23: W. S. C.	Blackley,	10 Hamilton Crescent, Partick.
1876, Jan. 25: John	Brown, B.Sc.,	11 Somerset Place, Glasgow.
1865, Jan. 18: John	Bryce,	Sweethope Cottage, N. Milton Road, Dunoon.

Names marked thus * were Associates of Scottish Shipbuilders' Association at incorporation with Institution, 1865.

1880, Dec. 21: John Cassells, 56 Cook Street, Glasgow.
1870, Dec. 20: Joseph J. Coleman, F.C.S., Ardarroch, Bearsden, by Glasgow.

1885, Feb. 24: Robert Darling, 5 Summerside Place, Leith.

1859, Nov. 23: Sir A. Orr Ewing, Bart., M.P., 2 W. Regent Street, Glasgow.

1885, Mar. 24: James S. Gardner, 52 North Frederick Street, Glasgow.
1863, Mar. 18: Robert Gardner, 52 North Frederick Street. Glasgow.

1860, Jan. 18: George T. Hendry, 79 Gt. Clyde St., Glasgow.

1882, Oct. 24: Wm. A. Kinghorn, 6 Colebrooke St., Hillhead, Glasgow.
1864, Dec. 21: Anderson Kirkwood,LL.D., 7 Melville Ter., Stirling.

1878, Oct. 29: John Langlands, 88 Gt. Clyde St., Glasgow.
1884, Feb. 26: C. R. Lemkes, 198 Hope Street, Glasgow.

1886, Jan. 26: Capt. Dun. M'Pherson, 142 Pollok Street, Glasgow.

1873, Feb. 18: John Mayer,F.C.S. 2 Clarinda Terrace, Pollokshields, Glasgow.

1874, Mar. 24: James B. Mercer, Broughton Copper Works, Manchester.

George *Miller, 1 Wellesley Place, Glasgow.
1865, Dec. 20: John Morgan, Springfield House, Bishopbriggs, Glasgow.

1883. Dec. 18: W. M'Ivor Morison, Mayfield, Marine Place, Rothesay.

James S. *Napier, 33 Oswald Street, Glasgow.

John Phillips, 17 Anderston Quay, Glasgow.

44

1869, Nov. 23: Capt. John Rankine,   31 Airlie Terrace, Pollok-shields, Glasgow.

1867, Dec. 11: William H. Richardson,   19 Kyle Street, Glasgow.

1882, Dec. 19: Colin Wm. Scott,   30 Buchanan St., Glasgow.

1876, Jan. 25: George   Smith,   45 West Nile St., Glasgow.

John   *Smith,   Aberdeen Steam Navigation Co., Aberdeen.

Malcolm M'N. *Walker,   45 Clyde Place, Glasgow.

H. J.   *Watson,   5 Oswald Street, Glasgow.

1882, Dec. 19: John D.   Young,   141 Buchanan St., Glasgow.

William   *Young,   Galbraith Street, Stobcross, Glasgow.

GRADUATES.

1884, Dec. 23: Arthur C. Auden,   ———

1882, Nov. 28: William H. Agnew,   Laird & Coy., Birkenhead.

1880, Nov. 2: James   Aitken,   2 Lawn Villas, Harringay Road, W. Green, Fottenham, London, N.

1885, Dec. 22: John Henry Alexander,   42 Sardinia Terrace, Hillhead, Glasgow.

1880, Feb. 24: George   Almond,   Belmont, Bolton-le-Moors, Lancashire.

1885, Dec. 22: Peter M'L. Baxter,   8 Mansfield Place, Blythswood Square, Glasgow.

1887, Apr. 26: Thomas   Bell,   Little Park Cottage, Yoker.

1885, Mar. 24: Alexander   Bishop,   3 Germiston St., Glasgow.

1885, Oct. 27: Archibald   Blair,   12 Arthur Street, Glasgow.

1883, Dec. 18: David   Blair,   Allan Line Works, Mavisbank Quay, Glasgow.

1884, Jan. 22: George	Blair, Jun.,	6 Alfred Terrace, Hillhead, Glasgow.
1884, Jan. 22: Henry	Blair,	Clutha Ironworks, Glasgow.
1885, Oct. 27: William C.	Borrowman,	15 Breadalbane Street, Glasgow.
1878, Dec. 17: Rowland	Brittain,	11 Mount Pleasant Road, Stroud Green, London, N.
1886, Oct. 26: James	Brown,	89 North Frederick Street, Glasgow.
1883, Apr. 24: Arthur R.	Brown,	5 Prince of Wales Terrace, Hillhead.
1879, Feb. 25: Alex. T.	Brown,	6 Olrig ;Terrace, Glencairn Drive, Pollokshields, Glasgow.
1883, Dec. 18: Eben. H.	Brown,	————
1885, Mar. 24: Matthew R.	Brown,	Little Park Cottage, Yoker.
1881, Jan. 25: Matthew T.	Brown, B.Sc.,	33 Hope Street, Glasgow.
1885, Dec. 22: Hugh	Brown,	Holmfauldhead, ! ' Renfrew Road, Govan.
1886, Oct. 26: J. B.	Buchanan,	175 Hope Street, Glasgow.
1876, Dec. 19: Lindsay	Burnet,	Moore Park Boiler Works, Govan, Glasgow.
1884, Feb. 26: John	Cleland, B.Sc.,	Woodhead Cottage, Old Monkland.
1881, Nov. 22: Alfred A. R.	Clinkskill,	1 Holland Place, Glasgow.
1884, Feb. 26: Alexander	Conner,	9 Scott Street, Glasgow.
1885, Dec. 22: Benjamin	Conner,	9 Scott Street, Glasgow.
1884, Jan. 22: Alex. M.	Copeland,	Bellahouston Farm, Paisley Road, Glasgow.
1880, Dec. 21: Sinclair	Couper,	6 Clayton Terrace,'Dennistoun, Glasgow.
1885, Oct. 27: Francis	Coutts,	96 Shamrock St., Glasgow.
1880, Nov. 23: James M.	Croom,	Ulster Ironworks, Abercorn Basin, Belfast.

1882, Feb. 21: Wm. S.　　Cumming,　　Blackhill, by Parkhead, Glasgow.

1884, Jan. 22: James　　Dalziel,　　20 Kelvinhaugh Street, Glasgow.

1886, Mar. 23: Thomas　　Danks,　　Burgh Chambers, Govan.

1883, Apr. 24: Alexander　Darling,　Upper Assam Tea Coy., Maijen Dilbrugdah,Upper Assam, India.

1881, Mar. 22: David　　Davidson,　19 Albert Drive, Crosshill, Glasgow.

1885, Feb. 24: William S.　Dawson,　Broomhill Iron Works, Glasgow.

1883, Dec. 18: William　　Denholm,　Hamilton Place, 370 Dumbarton Road, Glasgow.

1883, Feb. 10: Lewis M. T. Deveria,　Tharsis Huelva, Spain.

1886, Nov. 23: Thomas　　Dick,　　Bowling.

1882, Oct. 24: Daniel　　Douglas,　Earle's Shipbuilding Co., Hull.

1880, Nov. 2: Geo. C.　　Douglas,　Douglas Foundry, Dundee.

1882, Oct. 24: John F.　　Douglas,　18 Meadowpark Street, Dennistoun.

1883, Oct. 23: Harry W.　Downes,　8 South Crescent, Hartlepool.

1886, Nov. 23: George F.　Duncan,　Ardenclutha,Port-Glasgow.

1884, Jan. 22: William　　Dunlop,　953 Govan Road, Govan.

1885, Mar. 24: Robert　　Elliot, B.Sc., The Engineers' Club, 10 Hare Street, Calcutta.

1878, Jan. 22: James R.　　Faill,　　Craig-en-Callie, Ayr.

1882, Feb. 21: Albert E.　Fairman,　Woodlands, Garelochhead,

1880, Dec. 21: Henry M.　Fellows,　Westbourne Lodge, Great Yarmouth.

1884, Jan. 22: Thomas G.　Ferguson,　14 Queen's Cres., Glasgow.

1881, Feb. 22: William   Ferguson,   Larkfield, Partick.

1885, Jan. 27: Wm. D.   Ferguson,   37 Bentinck St., Glasgow.

1881, Nov. 22: Charles J.   Findlay,   10 Belmont Cres., Hillhead, Glasgow.

1883, Oct. 23: Duncan   Finlayson,   730 Govan Road, Govan, Glasgow.

1869, Oct. 26: F. P.   Fletcher,   South Russell St., Falkirk.

1886, Apr. 27: John I.   Fraser,   13 Sandyford Pl., Glasgow.

1885, Oct. 27: Henry G.   Gannaway,   17 Caroline Street, Jarrow-on-Tyne.

1886, Dec. 21: Charles H.   Geddes,   495 St. Vincent Street, Glasgow.

1874, Feb. 24: James   Gillespie,   21 Minerva St., Glasgow.

1884, Dec. 23: D. C.   Glen, Jun.,   14 Annfield Pl., Glasgow.

1885, Jan. 27: Alex. M.   Gordon,   2 Hawarden Place, Ibrox, Glasgow.

1882, Jan. 24: Arthur B.   Gowan,   3 Anderson Street, Port-Glasgow.

1884, Feb. 26: Alexander   Gracie,   9 Great George Street, Hill-head, Glasgow.

1887, Jan. 25: Edmund J.   Gumprecht,   Langdale, Dowanhill Gardens, Glasgow.

1881, Dec. 20: Andrew   Hamilton,   2 Belmar Terrace, Pollok-shields, Glasgow.

1881, Feb. 22: James   Harvey,   c/o Merry & Cunninghame, Glengarnock.

1883, Feb. 20: David   Henderson,   11 Hayburn Crescent, Par-tickhill, Glasgow.

1882, Nov. 28: F. N.   Henderson,   11 Princes Terrace, Dowan-hill, Glasgow.

1881, Oct. 25: Charles G.   Hepburn,   Ben Boyd Road, Neutral Bay, North Shore, Sydney, N.S.W., Australia.

1882, Feb. 21: Wm. S.	Herriot,	Leonora, Demerara.
1887, Feb. 22: George R.	Hingelberg,	48 Hairst Street, Renfrew.
1881, Jan. 25: A. C.	Holms, Jun.,	Hope Park, Partick, Glasgow.
1884, Dec. 23: John	Howarth,	37 Bentinck St., Glasgow.
1886, Oct. 26: Gilbert M.	Hunter,	3 Germiston St., Glasgow.
1883, Jan. 23: John A.	Inglis,	23 Park Circus, Glasgow.
1885, Feb. 24: John	Inglis,	Bonnington Brae, Edinburgh.
1873, Dec. 23: David	Johnston.	12 York Street, Glasgow.
1886, Nov. 23: Daniel	Kemp,	128 West Graham Street, Glasgow.
1883, Feb. 20: Eben. D.	Kemp,	Overbridge, Govan, Glasgow.
1886, Dec. 21: Donald	King,	6 Drummond St., Glasgow.
1886, Jan. 26: John	King,	8 Hamilton Street, Partick.
1885, Feb. 24: John	Lang,	5 Ruthven Street, Kelvinside, Glasgow.
1882, Jan. 24: Andrew	Laing,	2 Glenavon Terrace, Crow Road, Partick, Glasgow.
1886, Jan. 26: John	Lee,	Bleachfield House, Kilwinning.
1886, Dec. 21: Robert	Lee, Jun.,	2 Minard Terrace, Partick-hill, Glasgow.
1883, Nov. 21: William R.	Lester,.	2 Doune Terrace, North Woodside, Glasgow.
1885, Mar. 24: William	Linton,	————
1885, Mar. 24: Fred.	Lobnitz,	2 Park Terrace, Govan.
1884, Dec. 23: Robert	Logan,	3 Hayburn Cres., Partick, Glasgow.
1886, Dec. 21: William	Macclashan,	11 St. Paul's Square, Perth.

1880, Nov. 2: Patrick F.	M'Callum,	Fairbank Cottage, Helensburgh.
1881, Dec. 20: H.	M'Coll, Jun.,	3 Dalmeny Terrace, Pollokshields.
1883, Dec. 18: Peter	M'Coll,	Stewartville Place, Partick, Glasgow.
1883, Dec. 18: John	MacDonald,	293 New City Road, Glasgow.
1882, Oct. 24: James L.	Macfarlane,	Meadowbank, Torrance.
1887, Jan. 25: Dugald	M'Farlane,	54 Kelvingrove St., Glasgow.
1887, Jan. 25: David L.	M'Geachen,	56 Paterson Street, S.S.
1886, Dec. 21: John	Macgregor,	3 Germiston St., Glasgow.
1883, Dec. 18: John Bow	M'Gregor,	13 Clarendon St., Partick, Glasgow.
1886, Dec. 21: James	Mack,	3 Germiston St., Glasgow.
1880, Feb. 24: Neil	M'Kechnie,	8 Glenavon Terrace, Crow Road, Partick, Glasgow.
1881, Oct. 25: James	Mackenzie,	C/o D. Rollo & Sons, 10 Fulton Street, Liverpool.
1883, Jan. 23: Thos. B.	Mackenzie,	342 Duke Street, Glasgow.
1883, Feb. 26: Robert	M'Kinnell,	56 Dundas Street, S.S., Glasgow.
1883, Dec. 19: Colin D.,	M'Lachlan,	3 Rosehill Terrace, South-Queensferry.
1882, Dec. 19: Peter	M'Lean,	Waverley Ironworks, Galashiels.
1885, Jan. 27: John	M'Millan,	26 Ashton Ter., Glasgow.
1875, Dec. 21: Allister	M'Niven,	————
1886, Dec. 21: Andrew	M'Vitae,	385 Dumbarton Road, Glasgow.
1887, Feb. 22: Cree	Maitland,	4 Hampton Court Terrace, Renfrew Street, Glasgow.
1885, Oct. 27: John M.	Malloch,	Carron Iron Works, Falkirk.

1884, Dec. 23: Robert   Mansel, Jun., 4 Clyde View, Partick.

1884, Dec. 23: W. J.   Marshall,   3 Minerva Street, Glasgow.

1880, Nov. 2: Ivan   Mavor,   Wincomlee, Low Walker-on-Tyne.

1882, Jan. 24: Robt. Alex. Middleton,   20 Merryland St., Govan, Glasgow.

1884, Nov. 25: Thomas   Millar,   28 Wilberforce St., Wall-send-on-Tyne.

1880, Feb. 24: Robert   Miller,   13 Park Grove Terrace, W., Glasgow.

1883, Dec. 18: Charles W. Milne,   7 Carmichael Street, Govan.

1881, Jan. 25: Ernest W. Moir.   Forth Bridge Works, South Queensferry.

1882, Feb. 21: C. J.   Morch,   Horten, Norway.

1882, Nov. 28: M. J.   Morrison, B.Sc., 8 Annfield Ter., Partick, Glasgow.

1885, Dec. 22: William B. Morrison,   340 Dumbarton Road, Glasgow.

1884, Feb. 26: Andrew   Munro,   629 Govan Road, Govan, Glasgow.

1878, May 14: Angus   Murray,   4 Sutherland Ter., Dowanhill, Glasgow.

1883, Dec. 18: James L.   Napier,   808 St. Vincent Street, Glasgow.

1886, Jan. 26: Thomas   Nicholson,   6 Annfield Place, Glasgow.

1879, Nov. 25: Alex. R.   Paton,   Redthorn, Partick, Glasgow.

1886, Jan. 26: Samuel   Paxton,   4 Lorne Terrace, Pollokshields.

1873, Dec. 23: Edward C. Peck,   Yarrow & Co., Poplar, London, E.

1881, Oct. 25: William T. Philp,   284 Bath Street, Glasgow.

1887, Apr. 6: John C.	Preston,	27 Town Hall, Brisbane, Queensland.
1885, Jan. 27: James L.	Proudfoot,	154 West George Street, Glasgow.
1885, Feb. 24: John T.	Ramage,	The Hawthorn's, Bonnington, Edinburgh.
1886, Dec. 21: Andrew T.	Reid,	10 Woodside Terrace, Glasgow.
1883, Nov. 21: Hugh	Reid,	10 Woodside Terrace, Glasgow.
1884, Dec. 23: James G.	Reid, Jun.,	8 W. Princes St., Glasgow.
1886, Dec. 21: John	Reid,	10 Woodside Terrace, Glasgow.
1884, Feb. 26: Walter	Reid,	104 Armadale Street East, Glasgow.
1886, Oct. 26: Alexander	Robertson,	111 Kenmure Street, Pollokshields.
1886, Apr. 27: Robert	Robertson, B.Sc.,	154 West George St., Glasgow.
1882, Nov. 28: J. M'E.	Ross,	Ravensleigh, Dowanhill Gardens, Glasgow.
1885, Dec. 22: Chas. A. F.	Rubie,	———
1887, Apr. 6: Joseph W.	Russell,	461 St. Vincent St., Glasgow.
1884, Mar. 25: J. B.	Sanderson,	15 India Street, Glasgow.
1885, Oct. 27: Alexander	Scobie,	Culdees, Partickhill, Glasgow.
1879, Mar. 25: John	Scobie,	Culdees, Partickhill, Glasgow.
1886, Dec. 21: Walter	Scott,	28 Rowancraig Place, Glasgow Road, Dumbarton.
1886, Mar. 23: Thomas R.	Seath,	Sunny Oaks, Langbank.
1886, Mar. 23: William Y.	Seath.	Sunny Oaks, Langbank.

45

1880, Apr. 27: Archibald	Sharp,	City and Guilds of London Institute, Exhibition Rd., London, S.W.
1882, Oct. 24: John	Sharp,	461 St. Vincent St., Glasgow.
1853, Dec. 18: George	Simpson,	13 Maxwell Street, Partick, Glasgow.
1877, Mar 20: Nisbet	Sinclair, Jun.,	27 La Crosse Terrace, Hillhead, Glasgow.
1884, Mar. 25: Russell	Sinclair,	c/o J. R. C. Sinclair, 2 West Quay, Greenock.
1882, Nov. 28: Geo. H.,	Slight, Jun.,	84 George St., Edinburgh.
1881, Nov. 22: John A.	Steven,	12 Royal Crescent, Glasgow.
1881, Jan. 25: William	Stevenson,	R. & J. Hawthorn, St. Peter's Works, Newcastleon-Tyne.
1873, Dec. 23: John	Stewart,	270 New City Road, Glasgow.
1875, Dec. 21: Andrew	Stirling,	2 Greenvale Terrace, Dumbarton.
1884, Dec. 23: David W.	Sturrock,	11 Florence Pl., Glasgow.
1886, Dec. 21: James R.	Symington,	204 St. Vincent St., Glasgow.
1880, Dec. 21: Stanley	Tatham,	Northern Counties Club, Newcastle-on-Tyne.
1883, Dec. 18: Lewis	Taylor,	12 Hillsborough Terrace, Hillhead, Glasgow.
1882, Nov. 28: William	Taylor,	57 St. Vincent Cres., Glasgow.
1880, Nov. 23: George	Thomson,	120 Soho Hill, Handsworth, Birmingham.
1874, Feb. 24: George C.	Thomson,	39 Kersland Terrace, Hillhead, Glasgow.

1884, Dec. 23: John	Thomson,	15 Burnbank Gardens, Glasgow.
1884, Dec 23: William	Thomson,	15 Burnbank Gardens, Glasgow.
1885, Oct. 27: Peter	Tod,	2 Greenvale Terrace, Dumbarton.
1887, Jan. 25: David R.	Todd,	107 Hope Street, Glasgow.
1885, Feb 24: Charles H.	Wannop,	12 Derby Street, Glasgow.
1884, Feb. 26: William	Warrington,	23 Miller Street, Glasgow.
1881, Mar. 22: Robert	Watson,	1 Glencairn Drive, Pollokshields, Glasgow.
1880, Apr. 27: Robert D.	Watt,	s.s. "Wenchow," c/o Marine Engineering Institute, Shanghai, China.
1878, Dec. 17: Robert L.	Weighton, M.A.,	R. & J. Hawthorn, St. Peter's, Newcastle-on-Tyne.
1884, Apr. 22: John	Weir,	Ramage & Ferguson, Shipbuilders, Leith.
1886, Apr. 27: William	Weir,	1 Radbourne Terrace, Gateshead-on-Tyne.
1885, Nov. 24: James	Welsh,	51 St. Vincent Crescent Glasgow.
1882, Nov. 28: Geo. B.	Wemyss,	511 Springburn Rd., Glasgow.
1883, Dec. 18: John	Whitehead,	10 Witch Rd., Kilmarnock.
1877, Jan. 23: Robt. John	Wight,	7 Berlin Place, Pollokshields, Glasgow.
1886, Apr. 27: Percy F. C.	Willcox,	139 Allison St., Govanhill, Glasgow.
1879, Oct. 28: William	Willox, M.A.,	————
1883, Jan. 23: John	Wilson,	175 North Street, Glasgow
1883, Dec. 18: David	Wood,	124 West Nile Street, Glasgow.
1885, Mar. 24: Fred. W.	Zucker,	————●

CONTENTS

OF THE

TRANSACTIONS:

PAPERS, DISCUSSIONS, &c.

VOL. V.—FIFTH SESSION, 1861-62.

PROCEEDINGS

OF THE

Scottish Shipbuilders' Association.

INDEX.

PAGE

WILLIAM MUNRO, Law and General Printer, 80 Gordon Street, Glasgow.

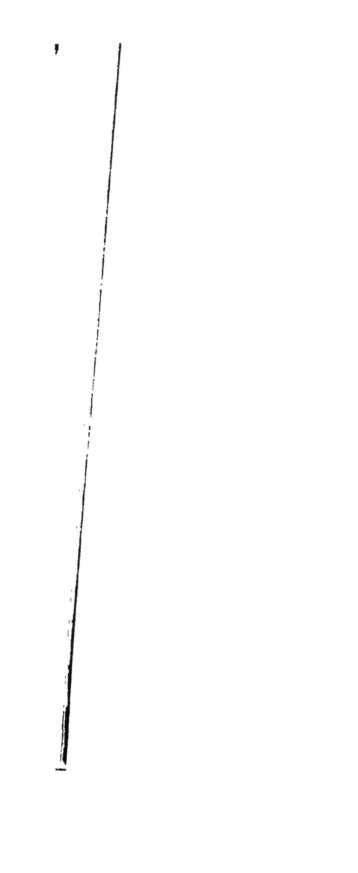

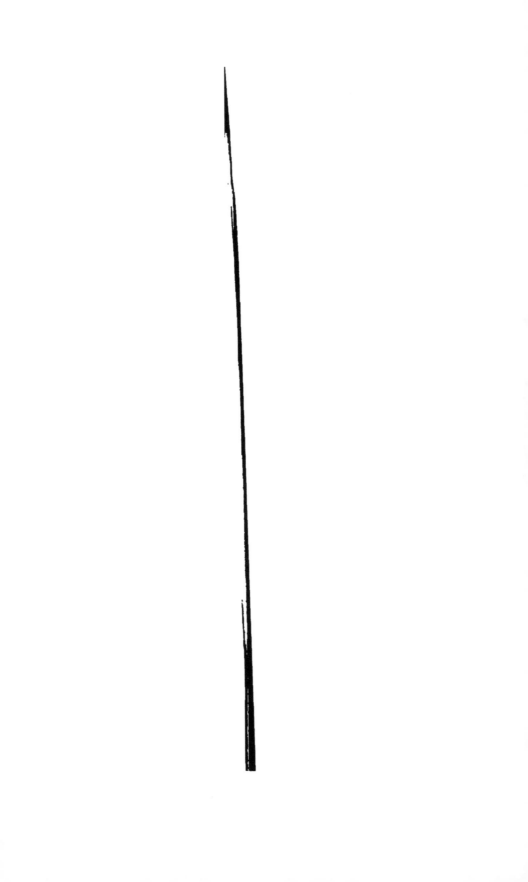

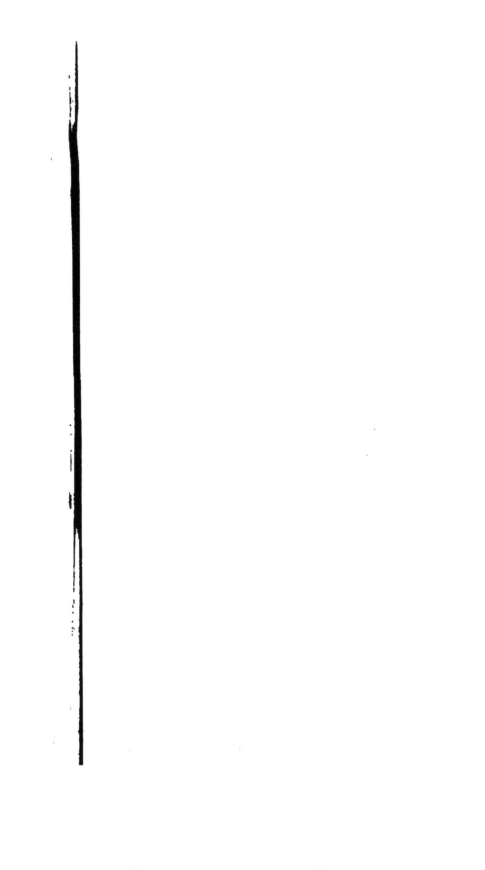

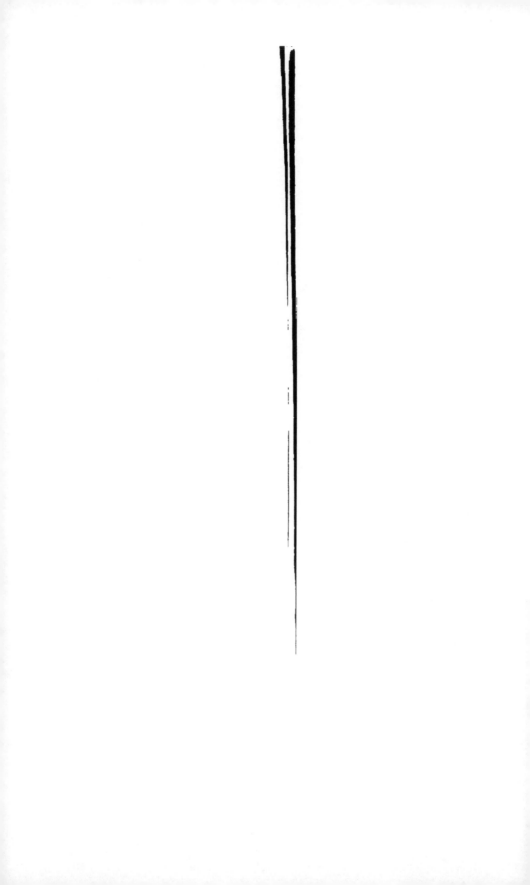

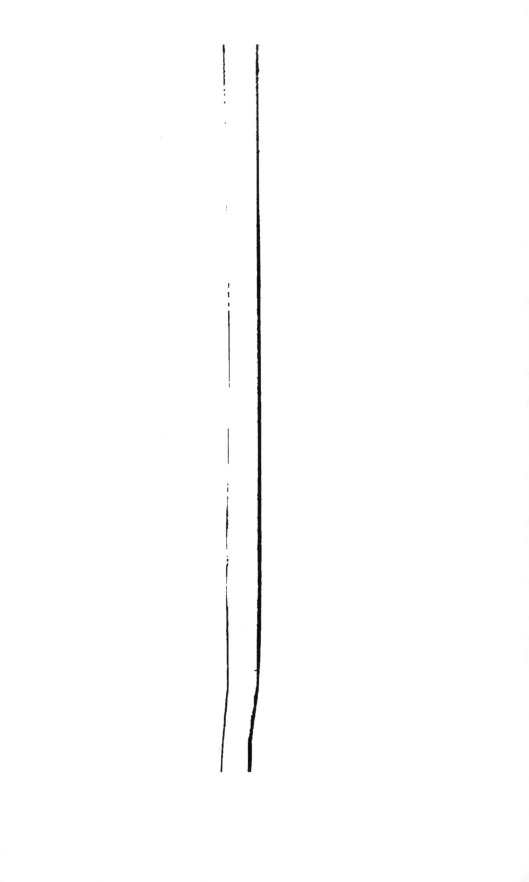